Introductory Algebra

by Chris Nord

with additional contributions
by Matthew Schmidgall and Martin Prather

Introductory Algebra
ISBN: 978-1-943536-56-6
Edition 1.0 Fall 2019
© 2019, Chemeketa Community College. All rights reserved.

Chemeketa Press

Chemeketa Press is a nonprofit publishing endeavor at Chemeketa Community College. Working together with faculty, staff, and students, we develop and publish affordable and effective alternatives to commercial textbooks. All proceeds from the sale of this book will be used to develop new textbooks. For more information, please visit www.chemeketapress.org.

Publisher: David Hallett
Director: Steve Richardson
Managing Editor: Brian Mosher
Manuscript Editors: Steve Richardson, Matt Schmidgall
Instructional Editor: Stephanie Lenox
Design Editors: Ronald Cox IV
Cover Design: Ronald Cox IV
Interior Design: Ronald Cox IV, Kristi Etzel
Layout: Ronald Cox IV, Matthew Sanchez, Brice Spreadbury

Chemeketa Math Faculty

The development of this text and its accompanying MyOpenMath classroom has benefited from the contributions of many Chemeketa math faculty in addition to the author, including:

Dan Barley, Lisa Healey, Kelsey Heater, Tim Merzenich, Nolan Mitchell, Chris Nord, Martin Prather, and Matthew Schmidgall.

Printed in the United States of America.

Contents

Chapter 1: Whole Numbers...1

   1.1 Addition and Subtraction.. 2

   1.2 Solving Equations...11

   1.3 Multiplication and Division..22

   1.4 Order of Operations...32

   1.5 Solving Equations.. 41

Chapter 2: Integers...**47**

   2.1 Introduction to Integers..48

   2.2 Addition and Subtraction 55

   2.3 Multiplication and Division...64

   2.4 Solving Equations 71

Chapter 3: Rational Numbers...**83**

   3.1 Introduction to Rational Numbers 84

   3.2 Divisibility 91

   3.3 Prime Factorization...100

   3.4 Equivalent Fractions... 107

   3.5 Multiplication and Division....................................... 115

   3.6 Comparing Rational Numbers 123

   3.7 Addition and Subtraction..130

   3.8 Solving Equations..136

Chapter 4: Expressions..**147**

   4.1 Introduction to Expressions.....................................148

   4.2 Combining Like Terms 161

   4.3 Simplifying Expressions with Algebraic Fractions.............171

   4.4 Modeling with Expressions and Equations 177

   4.5 Solving Equations..186

Chapter 5: Graphs and Tables..**195**

   5.1 Interpreting Graphical Representations of Data...............196

   5.2 Mean, Median, and Mode.......................................202

5.3 Equations with Two Variables...213

5.4 The Cartesian Plane...225

5.5 Graphs of Linear Equations...233

Chapter 6: Proportional Reasoning.. 249

6.1 Ratios and Rates..250

6.2 Solving Proportions 259

6.3 Dimensional Analysis 270

Solutions to Odd-Numbered Exercises 281

Glossary/Index 295

CHAPTER 1

Whole Numbers

What Are Numbers?

You know the answer to this question, of course, because you've been using numbers for most of your life. You use them every day. However, as you will see in the chapters to come, numbers can be a little more complicated than those things you use to count the money in your wallet or the people in front of you in the express lane at the grocery store. We will return to this question about what numbers are at the beginning of each chapter, and with each return, we'll add a new layer of sophistication to the answer.

To start, we'll just say that numbers are symbols that we use to count objects in a collection. This set of numbers is called the **whole numbers**. The whole numbers are {0, 1, 2, 3, 4, ...}. With the exception of zero, which has only been considered a number for about 1400 years, the set of whole numbers is as old as human civilization. In fact, there is some evidence of a surprisingly sophisticated innate understanding of numbers among several non-human species in the animal kingdom, too, so it may be that some version of the set of whole numbers *predates* humanity.

While there is a smallest whole number, zero, there is no greatest whole number. With any given whole number, there is always a next whole number that is greater than the given number.

In this chapter, we will explore the arithmetic of whole numbers and introduce the concept of an equation in the context of whole numbers. This chapter lays the foundation for the rest of the book, but it's important to remember that this chapter also rests on its own foundation — the ancient and basic concept of counting.

1.1 Addition and Subtraction

When we're presented with two collections of objects, we can count them separately, or we can combine them into a single collection and count them together. In this section, we will introduce the mathematical operation called *addition*, which encodes the relationship between these two ways of counting, and explore its properties.

If we wish to *remove* any objects from a collection of objects, we can either count the number of objects before any are removed and then count the number of objects that are removed, or we can count the number of objects that remain after we are finished removing objects. In this section, we will also introduce the mathematical operation called *subtraction*, which encodes the relationship between these two ways of counting, and explore its properties.

A. Addition

Whole numbers are the symbols we use to count the objects in a collection. So suppose that you have two collections of — *anything*. First you count the things in one collection. We'll use the letter a to represent that whole number. Then you count all the things in the second collection. We'll use the letter b to represent that whole number. If you combine these two collections into a single collection, you can figure out the total number of things you have by adding a and b.

Addition is the process of bringing two numbers together to make a new total. The math symbol for addition is the plus sign (+). The numbers being added, a and b, are called the **terms**, and the result of the addition is called the **sum**.

> ### Addition
>
> Addition is the operation where 2 numbers, let's call them a and b, are added together to create a sum, c. Symbolically, $a + b = c$.

For example, suppose that Ben and Joaquin go door to door in their neighborhoods to raise money for their baseball team. Ben secures 7 donations, and Joaquin brings his baby sister along to make himself look like a good brother and secures 11 donations. The combined number of donations that Ben and Joaquin secure is represented by the *sum* of the *terms* 7 and 11.

By the way, we use the word "sum" to mean the addition itself and also the result of the addition. So both of the following statements use "sum" correctly:

The sum of 7 and 11 is 7 + 11.

The sum of 7 and 11 is 18.

To **evaluate** the sum of two numbers is to calculate the result of their addition. If your teacher asks you to evaluate the sum of 7 and 11, then the "sum" you give your teacher is the result. "Eighteen," you say, or, "The sum of 7 and 11 is 18."

When we count Ben's and Joaquin's combined donations, it doesn't matter whether we count Ben's donations first and Joaquin's second. With 7 + 11, the sum is 18. With 11 + 7, the sum is also a combined total of 18 donations. This is true *whenever* we add two numbers together. Being able to add numbers in any order and get the same sum is a property of addition that we call the commutative property of addition.

Commutative Property of Addition

When a and b represent any two numbers, then $a + b = b + a$.

Addition as an operation only happens between two numbers at one time. In real life, however, we often use addition to combine three or more numbers. When we do this, we're actually calculating a series of two-number sums.

For example, suppose Luke is a teammate of Ben and Joaquin. Luke is kind of lazy, to be honest, and only secures 3 donations for the fundraiser. If we use addition to calculate the total number of donations secured by Ben, Joaquin, and Luke, we represent the total with the three-number sum of 7 + 11 + 3. To evaluate this sum, we can either add 7 and 11 first and then add 3 to the result of that sum, or we can get the same result by adding 7 to the sum of 11 and 3:

$(7 + 11) + 3 = 18 + 3 = 21$

$7 + (11 + 3) = 7 + 14 = 21$

This is true *whenever* we add three or more numbers together. It's another property of addition, and it's called the associative property of addition.

Associative Property of Addition

When a, b, and c represent any three numbers, then $(a + b) + c = a + (b + c)$.

The associative property of addition ensures that there is only one possible meaning for an expression such as $a + b + c$. Taken together, the commutative and associative properties of addition mean that we can add two or more numbers together in any order we please. The sum will always be the same no matter what the order is.

If we needed to add 7, 3, 5, and 4 together, we can write 7 + 3 + 5 + 4, or even change the order and write 3 + 4 + 5 + 7 and still get the same sum:

$7 + 3 + 5 + 4 = 19$

$3 + 4 + 5 + 7 = 19$

Since we know that we'll get the same sum regardless of the order numbers are added together, we can make evaluating addition problems easier. For example, to more easily evaluate 9 + 3 + 8 + 7 + 1, some people would mentally change the order of the terms to put numbers that add to 10 together:

$(9 + 1) + (3 + 7) + 8 = 10 + 10 + 8 = 28$

Example 1

Evaluate the following multi-term sums.

1. $8 + 4 + 2 + 7 + 6$

2. $13 + 9 + 2 + 21 + 7$

Solutions

1. $8 + 4 + 2 + 7 + 6$

   $8 + 4 + 2 + 7 + 6 = (8 + 2) + (4 + 6) + 7$ Re-order so that the terms adding to 10 are together.

   $\qquad\qquad\qquad = 10 + 10 + 7$

   $\qquad\qquad\qquad = 27$

2. $13 + 9 + 2 + 21 + 7$

   $13 + 9 + 2 + 21 + 7 = (13 + 7) + (21 + 9) + 2$ Re-order so that the terms adding to 20 or 30 are together.

   $\qquad\qquad\qquad = 20 + 30 + 2$

   $\qquad\qquad\qquad = 52$

Now let's go back to the story of the baseball team that is trying to raise money. Suppose that Carlos, the catcher, catches the chicken pox. Get it? He's the *catcher*, and he *catches* the chicken pox. Hilarious. Anyway, because he has chicken pox, Carlos can't go anywhere and doesn't secure *any* donations from his neighborhood. For Carlos, we use the whole number 0 to represent the number of donations that he secures. To calculate the combined number of donations that Carlos and Ben secure, we use the sum $0 + 7$. But the result of this addition is the same as the number of donations the Ben secured by himself, 7. We represent this with the following addition: $0 + 7 = 7$.

This same thing happens *whenever* we evaluate the sum of 0 and any other number. We call this the **identity property of addition**. We also call zero the **additive identity**.

Identity Property of Addition

When a represents any number, then:

$\qquad 0 + a = a$

$\qquad a + 0 = a.$

Let's practice using all three of these properties of addition.

Practice A

Use the commutative and associative properties of addition to reorder the following problems from smallest term to largest term and then evaluate. Identify which property you use for each step. Then turn the page to check your solutions.

1. $1 + 3 + 2$

2. $3 + 0$

3. $9 + 1 + 0$

4. $3 + 9 + 7$

5. $4 + 0 + 6 + 7$

6. $23 + 4 + 17$

B. Subtraction

When we remove objects from a collection, we use the binary operation **subtraction** to calculate the number of remaining objects. In mathematics, subtraction means taking something away from a group or number of things. When you subtract, the things in the group are fewer than before you subtracted. To describe how subtraction works, mathematics uses these technical terms:

The number of objects that is originally in the collection is called the **minuend**.

The number of objects you remove from the collection is called the **subtrahend**.

The number of objects that remain in the collection is called the **difference**.

The process of a subtraction is always **minuend** minus **subtrahend** equals **difference**. Using math symbols, that means $m - s = d$.

Suppose Simone, for example, is carrying a stack of 15 chocolate bars. Claire offers to help carry the chocolate bars, so Simone removes the top 6 chocolate bars from the stack and hands them to Claire. This leaves Simone with 9 chocolate bars to carry. In this example, the minuend is 15, the subtrahend is 6, and the difference is 9.

We can represent this example with the subtraction equation $15 - 6 = 9$. Just as you saw with the word "sum," we can use the word "difference" to mean the subtraction itself, $15 - 6$, or the result of the subtraction, 9. To *evaluate* a difference means to calculate the result of the difference.

So Simone now has 9 chocolate bars, and Claire has 6 chocolate bars. Notice that if we use addition to calculate the combined number of chocolate bars that Simone and Claire have, we end up with the same number of chocolate bars that Simone started off with, 15.

Every subtraction thus implies a related addition. In fact, we use the related addition as the basis for our definition of subtraction.

Subtraction

Let a represent a whole number, and let b represent a whole number that is less than or equal to a. The difference of a and b, $a - b$, is the whole number c, for which $a = b + c$.

In other words, $a - b = c$ if and only if $a = b + c$.

Example 2

Evaluate the following differences and then write the equivalent addition.

1. $15 - 6$ 2. $10 - 7$ 3. $22 - 14$

Solutions

1. $15 - 6 = 9$ We know the difference is 9 because of the equivalent addition from the definition of subtraction: $a - b = c$ if and only if $a = b + c$. $15 - 6 = 9$ because $15 = 6 + 9$.

2. $10 - 7 = 3$ because $10 = 7 + 3$.

3. $22 - 14 = 8$ because $22 = 8 + 14$.

Simone could have handed Claire more than 6 chocolate bars. She could have handed her 8 or 12 or even all 15 of the chocolate bars if she really wanted to take advantage of Claire's seemingly generous but perhaps sneaky offer. These scenarios would be represented by the differences $15 - 8$, $15 - 12$, and $15 - 15$.

However, it's impossible for Simone to give Claire more than 15 chocolate bars. She only has 15 chocolate bars, after all. When it comes to the whole numbers that we use to count things, a difference like $15 - 17$ doesn't make any sense. For now, then, we'll say that those kinds of differences are not whole numbers.

☠ Warning – Incorrect Approach! ☠

Addition Properties Do Not Apply to Subtraction

We saw earlier that the operation of addition has some important properties, particularly the commutative and associative properties. These properties allow us to change the order in which we add a string of numbers. However, please take note that these properties of addition do *not* apply to subtraction. When we subtract numbers, we must evaluate the two-number differences by moving from left to right. If we change the order of the numbers, we change the difference.

$$9 - 5 \neq 5 - 9$$

The difference on the left is 4, while the difference on the right is a negative integer. This new type of number will be explored in Chapter 2.

Example 3

Evaluate the following differences and sums from left to right.

1. $13 - 6 - 5$ 2. $10 - 3 + 4$

Solutions

1. $13 - 6 - 5$

$$13 - 6 - 5 = 7 - 5$$
$$= 2$$

If you ended up with the answer 12, it's because you evaluated $6 - 5$ first. With subtraction, you have to move from left to right, so the first difference on the left, $13 - 6$, has to be evaluated first.

2. $10 - 3 + 4$

$$10 - 3 + 4 = 7 + 4$$
$$= 11$$

If you ended up with the answer 3, it's because you evaluated the sum first.

Practice B

Now it's your turn to do some subtraction. Then turn the page to check your solutions.

7. $9 + 5 - 12$ 9. $15 - 7 + 2$ 11. $9 - 4 + 2 - 1$

8. $19 - 8 - 4$ 10. $12 - 7 - 1 + 3$ 12. $19 - 4 + 0 - 2$

Practice A — Answers

1.
$$1 + 3 + 2 = 1 + (3 + 2) \qquad \text{Apply the associative property.}$$
$$= 1 + (2 + 3) \qquad \text{Apply the commutative property.}$$
$$= 1 + 5 \qquad \text{Simplify by adding } 2 + 3.$$
$$= 6 \qquad \text{Simplify by adding } 1 + 5.$$

2.
$$3 + 0 = 0 + 3 \qquad \text{Apply the commutative property.}$$
$$= 3 \qquad \text{Apply the identity property.}$$

3. $9 + 1 + 0 = (9 + 1) + 0$ Apply the associative property.

$= (1 + 9) + 0$ Apply the commutative property.

$= 1 + (9 + 0)$ Apply the associative property.

$= 1 + (0 + 9)$ Apply the commutative property.

$= (1 + 0) + 9$ Apply the associative property.

$= (0 + 1) + 9$ Apply the commutative property.

$= 1 + 9$ Apply the identity property.

$= 10$ Simplify the sum.

4. $3 + 9 + 7 = 3 + (9 + 7)$ Apply the associative property.

$= 3 + (7 + 9)$ Apply the commutative property.

$= (3 + 7) + 9$ Apply the associative property.

$= 10 + 9$ Simplify by adding $3 + 7$.

$= 19$ Simplify by adding $10 + 9$.

5. $4 + 0 + 6 + 7 = (4 + 0) + 6 + 7$ Apply the associative property.

$= (0 + 4) + 6 + 7$ Apply the commutative property.

$= 4 + 6 + 7$ Apply the identity property.

$= 10 + 7$ Simplify by adding $4 + 6$.

$= 17$ Simplify by adding $10 + 7$.

6. $23 + 4 + 17 = (23 + 4) + 17$ Apply the associative property.

$= (4 + 23) + 17$ Apply the commutative property.

$= 4 + (23 + 17)$ Apply the associative property.

$= 4 + (17 + 23)$ Apply the commutative property.

$= 4 + 40$ Simplify by adding $17 + 23$.

$= 44$ Simplify by adding $4 + 40$.

Practice B — Answers

7. $9 + 5 - 12 = 14 - 12$ Simplify by adding 9 + 5.

$= 2$ Simplify by subtracting 14 − 12.

8. $19 - 8 - 4 = 11 - 4$ Simplify by subtracting 19 − 8.

$= 7$ Simplify by subtracting 11 − 4.

9. $15 - 7 + 2 = 8 + 2$ Simplify by subtracting 15 − 7.

$= 10$ Simplify by adding 8 + 2.

10. $12 - 7 - 1 + 3 = 5 - 1 + 3$ Simplify by subtracting 12 − 7.

$= 4 + 3$ Simplify by subtracting 5 − 1.

$= 7$ Simplify by adding 4 + 3.

11. $9 - 4 + 2 - 1 = 5 + 2 - 1$ Simplify by subtracting 9 − 4.

$= 7 - 1$ Simplify by adding 5 + 2.

$= 6$ Simplify by subtracting 7 − 1.

12. $19 - 4 + 0 - 2 = 15 + 0 - 2$ Simplify by subtracting 19 − 4.

$= 15 - 2$ Apply the additive identity property

$= 13$ Simplify by subtracting 15 − 2.

Exercises 1.1

For the following exercises, evaluate the sums.

1.   8 + 3

2.   5 + 9

3.   3 + 8

4.   9 + 5

5.   14 + 45

6.   63 + 25

7.   85 + 47

8.   56 + 78

9.   472 + 596

10.   638 + 295

11.   Which property guarantees that the answers for problems 1 and 3 are the same?

12.   Which property guarantees that the answers for problems 2 and 4 are the same?

For the following exercises, evaluate the three-number sums by adding the numbers in parentheses first.

13. $(5 + 9) + 11$

14. $7 + (23 + 10)$

15. $5 + (9 + 11)$

16. $(7 + 23) + 10$

17. $(8 + 6) + 40$

18. $(5 + 1) + 24$

19. $79 + (88 + 89)$

20. $77 + (40 + 92)$

21. Which property guarantees that the answers for problems 13 and 15 are the same?

22. Which property guarantees that the answers for problems 14 and 16 are the same?

For the following exercises, evaluate the multi-number sums. Don't forget that you can change the order of the numbers to make your work easier.

23. $2 + 15 + 8 + 5$

24. $3 + 9 + 27 + 1$

25. $17 + 36 + 23 + 15 + 4$

26. $42 + 21 + 8 + 32 + 19$

27. $77 + 49 + 13 + 15 + 21$

28. $15 + 36 + 15 + 17 + 54$

29. $82 + 65 + 35 + 18 + 7$

30. $87 + 38 + 12 + 13 + 5$

For the following exercises, write out the addition that is equivalent to each subtraction.

31. $36 - 25 = 11$

32. $52 - 14 = 38$

33. $478 - 261 = 217$

34. $523 - 96 = 427$

35. $887 - 38 = 849$

36. $248 - 69 = 179$

37. $7611 - 4653 = 2958$

38. $7024 - 6687 = 337$

For the following exercises, evaluate the differences. If the difference is not a whole number, say so and move on.

39. $9 - 4$

40. $7 - 3$

41. $12 - 12$

42. $13 - 19$

43. $0 - 15$

44. $73 - 29$

45. $64 - 48$

46. $159 - 35$

47. $476 - 132$

48. $37 - 0$

For the following exercises, evaluate the sums and differences.

49. $9 - 6 + 2$

50. $8 - 4 + 3$

51. $5 + 4 - 7 + 3$

52. $8 + 7 - 3 + 2$

53. $23 - 11 - 8$

54. $48 - 27 - 19$

55. $70 - 26 - 19 + 3$

56. $89 - 32 + 24 + 3$

1.2 Solving Equations

In this section, we introduce one of the most important concepts in this course — equations. We learn what an equation is and what it means to solve an equation. We then learn some basic techniques for solving equations and confirming that a solution is correct.

A. Solutions of Equations

An **equation** is a statement that two mathematical expressions have the same value. It's important to keep in mind that such a statement might be true, false, or incomplete and still be an equation.

Here are some examples of equations:

$7 + 13 = 20$ This equation is a *true* statement because the sum of 7 and 13 is 20.

$5 + 13 = 20$ This equation is a *false* statement because the sum of 5 and 13 is 18, not 20.

$x + 13 = 20$ This equation is *incomplete*. The letter x could represent any number value, so this equation has the potential to be a true statement, but it also has the potential to be a false statement.

In the last equation above, the letter x is called a **variable**. Variables are symbols — usually in the form of a single letter — that can represent any number value.

When we replace a variable symbol with a number value, we assign that value to the variable. We can then tell whether the resulting equation is a true or false statement. If we assign the value 5 to the variable x, for example, the incomplete equation $x + 13 = 20$ becomes the false statement $5 + 13 = 20$. On the other hand, if we assign the value 7 to the variable x, then the incomplete equation $x + 13 = 20$ becomes the true statement $7 + 13 = 20$.

If an equation has a variable, then any value we assign to the variable that makes the equation a true statement is called a **solution** of the equation. Solving an equation means finding all of the possible solutions to that equation.

▸ **Example 1**

In the following equations, show that the given value of the variable *is* or *is not* a solution to the given equation.

  1. Equation: $17 + t = 29$; variable value: $t = 12$

  2. Equation: $16 - m = 9$; variable value: $m = 5$

Solutions

1. $t = 12$ *is* a solution of $17 + t = 29$ because $17 + 12 = 29$ is a *true* statement.

2. $m = 5$ *is not* a solution of $16 - m = 9$ because $16 - 5 = 9$ is a *false* statement.

Now it's your turn. Figure out whether the following numbers are solutions of the given equations.

Practice A

Show that the given value of the variable *is* or *is not* a solution of the given equation. Then turn the page to check your solutions.

1. Equation: $13 + x = 22$; variable value: $x = 19$
2. Equation: $37 - d = 12$; variable value: $d = 25$
3. Equation: $b - 15 = 7$; variable value: $b = 8$
4. Equation: $22 + p = 26$; variable value: $p = 4$
5. Equation: $6 + k = 13$; variable value: $k = 7$

B. Using Equivalent Equations to Solve Equations

Take a minute to look at the following three equations. Which of the three is the easiest to solve? Which is the hardest to solve?

$$21 = 13 + x$$
$$21 - 13 = x$$
$$8 = x$$

The third equation is the easiest to solve. In fact, there's nothing that needs to be done! We simply read the solution from the equation. The only solution for x is the number value 8. Any other value that we assign to the variable will result in a false statement.

Equations like this third one, *variable = number* or *number = variable*, are called **assignment statements**. We use equations like this to assign a number value to a variable. When we solve an equation, we write the solution as an assignment statement.

Now compare the first and second equations. Wouldn't you agree that the first equation is a bit harder to solve than the second? To solve the second equation, all that we need to do is evaluate the difference of 21 and 13. If x is assigned any value other than the difference of 21 and 13, then the equation will be a false statement. So here is the solution of the second equation:

$$21 - 13 = x$$
$$8 = x$$

It turns out that the first and second equations are equivalent to each other. $21 - 13 = x$ is the same as $21 = 13 + x$. These two equations are equivalent because of the **definition of subtraction** which states $a - b = c$ if and only if $a = b + c$.

Let's see how this works with this second equation. If we replace a with 21, b with 13, and c with x, then the equations $a - b = c$ and $a = b + c$ become $21 - 13 = x$ and $21 = 13 + x$.

The definition of subtraction tells us that any number value that we assign to x that makes *one* of these equations a true statement must also make the *other* equation a true statement. That's what it means for equations like $21 - 13 = x$ and $21 = 13 + x$ to be **equivalent equations**.

Equivalent Equations

Two equations are equivalent if every solution of one equation is also a solution of the other equation.

To solve an equation, we write a sequence of equivalent equations. The last of those equivalent equations is an assignment statement equation. That assignment statement is the solution of the equation.

Example 2

Solve the following equations.

1. $35 = 24 + x$

2. $16 = y + 5$

Solutions

1. $35 = 24 + x$

   $35 - 24 = x$ This equation is equivalent to the first one because of the definition of subtraction.

   $11 = x$ This equation is equivalent to the second one because the difference of 35 and 24, $35 - 24$, is equal to the number 11.

2. $16 = y + 5$

   $16 - 5 = y$ The first two equations are equivalent because of the definition of subtraction.

   $11 = y$ The second and third equations are equivalent because the difference of 16 and 5, $16 - 5$, is equal to the number 11.

Now it's your turn to solve some equations using equivalent equations.

Practice B

Solve the following with equivalent equations. Then turn the page to check your solutions.

6. $81 = 27 + k$ 8. $45 = n + 18$ 10. $77 = 27 + b$

7. $33 = 19 + h$ 9. $43 = a + 35$ 11. $64 = g + 37$

C. Fact Families

Every addition or subtraction fact is part of a family of four equivalent equations called an **addition/subtraction fact family**.

   The previous examples illustrate how every subtraction fact corresponds to an addition fact through the definition of subtraction. These corresponding facts are equivalent equations that are part of the fact family. Because addition is commutative, we also know that each addition fact can be written in two different ways. These are also equivalent equations and part of the fact family.

Addition/Subtraction Fact Family

Assigning number values or variables to a, b and c in the following four equations results in an addition/subtraction fact family.

1. $a - b = c$ 2. $a = b + c$ 3. $a = c + b$ 4. $a - c = b$

Equations 1 and 2 above are equivalent to each other because of the *definition of subtraction*. Equations 2 and 3 are equivalent to each other because of the *commutative property of addition*. Equations 3 and 4 are equivalent to each other because of the *definition of subtraction*.

   We can also write an equation that is equivalent to any given equation by moving the expression on the left side of the equals sign to the right side and moving the expression on the right side to the left. This is called the **symmetric property of equality**.

Symmetric Property of Equality

If A and B represent expressions, then $A = B$ and $B = A$ must be equivalent equations.

Keep the symmetric property of equality in mind when working with fact families. It can be used to increase the number of equations in the fact family to 8. However, $a - b = c$ and $c = a - b$ will be treated as the same equation.

Example 3

Write the complete addition/subtraction fact family for the following equations.

1. $13 - 9 = z$ 2. $w - 4 = 20$

Solutions

1. If we use the first equation in the fact family, $a - b = c$, where $a = 13$, $b = 9$, and $c = z$, the equation numbers below correspond to the ones in the fact family definition box.

   a. $13 - 9 = z$ c. $13 = z + 9$

   b. $13 = 9 + z$ d. $13 - z = 9$

2. The equation $w - 4 = 20$ has the same form as equation 4 from the fact family definition box. Let's use equation 4, $a - c = b$, to come up with this fact family of equivalent equations.

   a. $w - 20 = 4$ **c.** $w = 4 + 20$

▶ **b.** $w = 20 + 4$ **d.** $w - 4 = 20$

Practice C

Write the complete addition/subtraction fact family for the following equations. Then turn the page to check your solutions.

12. $25 - x = 16$ **14.** $14 = 19 - y$ **16.** $n = 74 + 13$

13. $p + 6 = 31$ **15.** $34 = r + 7$ **17.** $21 = 47 - w$

Practice A — Answers

1. $x = 19$ *is not* a solution of $13 + x = 22$ because $13 + 19 = 22$ is *false*.
2. $d = 25$ *is* a solution of $37 - d = 12$ because $37 - 25 = 12$ is *true*.
3. $h = 8$ *is not* a solution of $h - 15 = 7$ because $8 - 15 = 7$ is *false*.
4. $p = 4$ *is* a solution of $22 + p = 26$ because $22 + 4 = 26$ is *true*.
5. $k = 7$ *is* a solution of $6 + k = 13$ because $6 + 7 = 13$ is *true*.

Practice B — Answers

6.
$$81 = 27 + k$$
$$81 - 27 = k$$
$$54 = k$$

9.
$$43 = a + 35$$
$$43 - 35 = a$$
$$8 = a$$

7.
$$33 = 19 + h$$
$$33 - 19 = h$$
$$14 = h$$

10.
$$77 = 27 + b$$
$$77 - 27 = b$$
$$50 = b$$

8.
$$45 = n + 18$$
$$45 - 18 = n$$
$$27 = n$$

11.
$$64 = g + 37$$
$$64 - 37 = g$$
$$27 = g$$

D. Solving Equations with Fact Families

To solve equations like $w = 7 + 9$ and $13 - 7 = t$, we evaluate the required sum or difference. With $w = 7 + 9$, we add $7 + 9$ to arrive at 16, so $w = 16$. With $13 - 7 = t$, we subtract 7 from 13 to arrive at 6, so $6 = t$. These solutions are assignment statements. The variable is alone on one side of the equal sign, and a sum or a difference is on the other side.

To solve an equation like $k - 14 = 23$ or $8 + z = 15$, where the variable is a part of a sum or difference, we can look at the addition/subtraction fact family for that equation and find an equation that has the variable alone on one side of the equation. Then we solve that equation from there.

Example 4

Use fact families to solve the following equations.

1. $k - 14 = 23$ 2. $8 + z = 15$

Solutions

1. $k - 14 = 23$

$k = 14 + 23$ We find an equivalent equation from the fact family with k on the left side of the equals sign.

$k = 37$ Now we evaluate the sum on the right side to find the solution.

2. $8 + z = 15$

$z = 15 - 8$ We find an equivalent equation from the fact family with z on the left side of the equals sign.

$z = 7$ Now we evaluate the difference on the right side to find the solution.

Practice D

Use fact families to solve the following equations. Then turn the page to check your solutions.

18. $13 + y = 17$ 20. $16 = r + 3$ 22. $64 - e = 3$

19. $t - 4 = 22$ 21. $4 = 13 - w$ 23. $k + 13 = 25$

Practice C — Answers

12. The equation $25 - x = 16$ has the same format as equations 1 and 4 of the fact family definition box. If we use the first equation of the fact family, $a - b = c$, where $a = 25$, $b = x$, $c = 16$, the equation numbers below correspond to the ones in the fact family definition box.

a. $25 - x = 16$ c. $25 = 16 + x$

b. $25 = x + 16$ d. $25 - 16 = x$

13. The equation $p + 6 = 31$ has the same format as equations 2 and 3 of the fact family definition box. If we use the second equation in the fact family, $a = b + c$, where $a = 31$, $b = p$, and $c = 6$, the equation numbers below correspond to the ones in the fact family definition box.

a. $31 - 6 = p$ c. $31 = p + 6$

b. $31 = 6 + p$ d. $31 - p = 6$

14. The equation $14 = 19 - y$ has the same format as equations 1 and 4 of the fact family definition box. If we use the fourth equation in the fact family, $a - c = b$, where $a = 19$, $b = 14$, and $c = y$, the equation numbers below correspond to the ones in the fact family definition box.

a. $19 - 14 = y$ c. $19 = y + 14$

b. $19 = 14 + y$ d. $19 - y = 14$

15. The equation $34 = r + 7$ has the same format as equation 2 and 3 of the fact family definition box. if we used the third equation in the fact family, $a = c + b$, where $a = 34$, $b = 7$, and $c = r$, the equation numbers below correspond to the ones in the fact family definition box.

a. $34 - 7 = r$ c. $34 = r + 7$

b. $34 = 7 + r$ d. $34 - r = 7$

16. The equation $n = 74 + 13$ has the same format as equations 2 and 3 of the fact family definition box. If we use the second equation in the fact family, $a = b + c$, where $a = n$, $b = 74$, and $c = 13$, the equation numbers below correspond to the ones in the fact family definition box.

a. $n - 74 = 13$ c. $n = 13 + 74$

b. $n = 74 + 13$ d. $n - 13 = 74$

17. The equation $21 = 47 - w$ has the same format as equations 1 and 4 of the fact family definition box. if we use the first equation in the fact family, $a - b = c$, where $a = 47$, $b = w$, and $c = 21$, the equation numbers below correspond to the ones in the fact family definition box.

a. $47 - w = 21$ c. $47 = 21 + w$

b. $47 = w + 21$ d. $47 - 21 = w$

E. Checking Your Solutions

In this section, we've learned how to solve certain types of equations, and we've also learned how to determine whether a given number is or is not a solution of an equation. However, we're not quite done.

Because no one is perfect and mistakes can happen, we should always check our solutions when we're asked to solve an equation. We do this by replacing the variable in the original problem with our solution. If that value creates a true statement, then our solution is truly a solution.

Example 5

Solve the following equations and then check the solution.

1. $18 = p + 7$ **2.** $9 - r = 14$

Solutions

1. First we solve the equation.

$$18 = p + 7$$

$18 - 7 = p$ We find an equivalent equation from the fact family with p isolated on the right side of the equals sign.

$11 = p$ Now we subtract the difference on the left side to find the solution.

We check the solution by substituting 11 for p in the original equation.

$$18 = 11 + 7$$
$$18 = 18$$

The statement $18 = 18$ is true, so we've verified the solution.

2. Now we solve the second equation.

$$9 - r = 14$$
$$r = 14 - 9$$
$$r = 5$$

Time to check the solution by substituting 5 for r in the original equation.

$$9 - 5 = 14$$
$$4 = 14$$ Uh oh.

Hmm. 4 does *not* equal 14, written $4 \neq 14$, so we have a false statement. That means we've made a mistake when we were solving this equation.

Can you spot the error in the solution for number 2? If we write out the full addition/subtraction fact family for $9 - r = 14$, we see that $r = 14 - 9$ is not actually in the same fact family as $9 - r = 14$. We will find the solution for this type of equation in Chapter 2.

It's important to develop the habit of checking your solutions. It helps you search for and correct any errors you may have made along the way, but more importantly, it helps you become a more independent and confident math student.

Practice E

Solve the following equations and check your solutions on your own. Then turn the page to check your solutions.

24. $17 + d = 20$

25. $s - 15 = 6$

26. $12 = q + 2$

27. $25 - p = 16$

28. $x + 27 = 43$

29. $59 = n - 24$

Exercises 1.2

For the following exercises, determine whether the variable value is a solution of the equation.

1. Equation: $x - 7 = 1$; variable value: $x = 8$

2. Equation: $9 = t - 5$; variable value: $13 = t$

3. Equation: $7 = b - 3$; variable value: $4 = b$

4. Equation: $x - 8 = 5$; variable value: $x = 3$

5. Equation: $25 - m = 18$; variable value: $m = 7$

6. Equation: $42 - c = 23$; variable value: $c = 19$

7. Equation: $31 = 16 + w$; variable value: $25 = w$

8. Equation: $77 = 14 + x$; variable value: $53 = x$

9. Equation: $274 + v = 625$; variable value: $351 = v$

10. Equation: $562 = 859 - q$; variable value: $q = 297$

Practice D — Answers

18. $13 + y = 17$

$$y = 17 - 13$$

$$y = 4$$

19. $t - 4 = 22$

$$t = 22 + 4$$

$$t = 26$$

20. $16 = r + 3$

$$16 - 3 = r$$

$$13 = r$$

21. $4 = 13 - w$

$$13 - 4 = w$$

$$9 = w$$

22. $64 - e = 3$

$$64 - 3 = e$$

$$61 = e$$

23. $k + 13 = 25$

$$25 - 13 = k$$

$$12 = k$$

Solve the following equations by evaluating a sum or difference.

11. $h = 6 + 13$

12. $24 - 14 = t$

13. $y = 19 - 14$

14. $35 + 27 = k$

15. $32 - 30 = x$

16. $m = 10 + 18$

17. $396 + 362 = w$

18. $r = 862 - 101$

19. $x = 364 - 97$

20. $75 + 287 = p$

For the following exercises, write out the full addition/subtraction fact family for each equation.

21. $v - 8 = 13$

22. $11 + q = 24$

23. $16 = r + 7$

24. $17 = 32 - t$

25. $25 - x = 9$

26. $34 = 19 + m$

27. $s + 85 = 113$

28. $57 = w - 73$

Practice E — Answers

24.
$$17 + d = 20$$
$$d = 20 - 7$$
$$d = 3$$
Check the solution by substituting 3 for d in the original equation.
$$17 + 3 = 20$$
$$20 = 20 \checkmark$$

25.
$$s - 15 = 6$$
$$s = 6 + 15$$
$$s = 21$$
Check the solution by substituting 21 for s in the original equation.
$$21 - 15 = 6$$
$$6 = 6 \checkmark$$

26.
$$12 = q + 2$$
$$12 - 2 = q$$
$$10 = q$$
Check the solution by substituting 10 for q in the original equation.
$$12 = 10 + 2$$
$$12 = 12 \checkmark$$

27.
$$25 - p = 16$$
$$p = 25 - 16$$
$$p = 9$$
Check the solution by substituting 9 for p in the original equation.
$$25 - 9 = 16$$
$$16 = 16 \checkmark$$

28.
$$x + 27 = 43$$
$$x = 43 - 27$$
$$x = 16$$
Check the solution by substituting 16 for x in the original equation.
$$16 + 27 = 43$$
$$43 = 43 \checkmark$$

29.
$$59 = n - 24$$
$$59 + 24 = n$$
$$83 = n$$
Check the solution by substituting 83 for n in the original equation.
$$59 = 83 - 24$$
$$59 = 59 \checkmark$$

For the following exercises, use fact families to solve the equations. Be sure to check your solution.

29. $v - 8 = 13$

30. $11 + q = 24$

31. $16 = r + 7$

32. $17 = 32 - t$

33. $25 - x = 9$

34. $34 = 19 + m$

35. $s + 85 = 113$

36. $57 = w - 73$

37. $528 + n = 765$

38. $z - 468 = 374$

39. $33 = x - 4$

40. $v + 10 = 18$

41. $68 = 33 + q$

42. $82 - k = 57$

43. $52 = 87 - y$

44. $94 = v + 87$

45. $90 = 345 - w$

46. $h - 413 = 143$

47. $420 = b - 83$

48. $446 + r = 720$

49. $116 - g = 80$

50. $865 = 928 - d$

1.3 Multiplication and Division

In this section, we introduce the operations of multiplication and division and describe their properties. We also introduce exponent notation and area of a rectangle. After completing this section the student should be able to:

- ◆ Evaluate and interpret products of whole numbers
- ◆ Use properties of multiplication as appropriate
- ◆ Evaluate and interpret whole number powers
- ◆ Evaluate and interpret quotients of whole numbers
- ◆ Evaluate a mixed sequence of multiplications and divisions
- ◆ Evaluate and interpret products, powers, and quotients involving zero

A. Multiplication

William works in a college cafeteria, where he regularly loads drinking glasses into a dishwasher rack. It's not a great job, but he likes the people and the opportunity to do mental math.

On Tuesday, William decides to figure out how many racks it will take to wash all of the glasses that need to be washed. The first step is to find out how many glasses fit in one rack. He can simply count the individual slots in the rack, but he's a bit of a mental math wizard and knows a more efficient counting method. Because the slots for glasses are arranged in rows, and because each row has five slots, William counts the slots in groups of five. The 6 rows of 5 add up to a total of 30 slots.

This type of counting, **skip counting**, is the basis for the operation called multiplication.

Figure 1

Multiplication

Let a and b represent any two whole numbers. The multiplication of a and b is the result of the sum $b + b + \ldots + b$ in which the number of terms is a. In math symbols:

$$\underbrace{b + b + \ldots + b}_{\text{There are } a \text{ terms.}} = a \cdot b$$

For example, $4 \cdot 3$ and $3 \cdot 4$ can be written in the following ways:

$4 \cdot 3 = 3 + 3 + 3 + 3$

$3 \cdot 4 = 4 + 4 + 4$

The numbers that go into the multiplication, a and b, are called the **factors**, and the result of the multiplication is called the **product**. The multiplication itself is also called the **product** of a and b. So the product of 5 and 6 is $5 \cdot 6$, and the product of 5 and 6 is also 30.

The math symbol $a \cdot b$ can be read as any of the following:

the product of a and b

the multiplication of a and b

a multiplied by b

a times b

There are also several different symbols that may be used for multiplication. Each of the following mean exactly the same thing:

$a \cdot b$

$a \times b$

$a * b$

$(a)(b)$

ab

In the last option, there's no operation symbol at all! However, this only works if at least one of the factors is either in parentheses or a variable. It wouldn't make a lot of sense to interpret 23 to be the product of 2 and 3 because it is the number 23. Similarly, we should be cautious when using the form $a \times b$ because the letter x is often used as a variable.

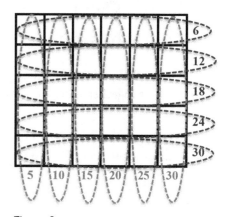

Because the drinking glass slots in William's dishwasher rack are arranged in a honeycomb pattern, the rows going the other direction are not straight, but William is smart and knows we can make a sketch of what the rack would look like if the slots were arranged in a rectangular array instead. In this sketch, we see that we have two options for skip counting the number of drinking glass slots:

Figure 2

We can count off 5 rows of 6 slots: $5 \cdot 6 = 30$

We can count off 6 rows of 5 slots: $6 \cdot 5 = 30$

This illustrates an important property, the **commutative property of multiplication**.

Commutative Property of Multiplication

When a and b are any two numbers, then $a \cdot b = b \cdot a$.

The rearranged tumbler rack in Figure 2 leads us to another important insight into how multiplication works. Suppose that William's diagram represented a rectangle measuring 6 feet long and 5 feet wide. Instead of drinking glass slots, the squares that are packed into the rectangle are 1 foot by 1 foot area units. Each square represents 1 **square foot** of area. The number of square feet that can be packed into the rectangle without any of them overlapping is the **area** of the rectangle.

We see from Williams' chart that the area of the 6 by 5 foot rectangle is the product of 6 and 5.

Area of a Rectangle

A rectangle measuring l units long and w units wide has area $l \cdot w$ square units.

Getting back to William and his dishwasher trays, suppose that he has four of these trays. He could count the total number of slots in several different ways. Here are two:

Count off 4 trays of 30 slots each: $4 \cdot (5 \cdot 6) = 4 \cdot 30 = 120$

Count off 20 rows of 6 slots each: $(4 \cdot 5) \cdot 6 = 20 \cdot 6 = 120$

This illustrates another important property, the **associative property of multiplication**.

Associative Property of Multiplication

When a, b and c are any three numbers, then $a \cdot (b \cdot c) = (a \cdot b) \cdot c$

We saw that addition has these same properties. Just like addition, when multiplying more than two numbers together, we can re-order the factors any way we want and still arrive at the same result.

Example 1

Evaluate the product of 2, 3, 5 and 7 in the given orders

1. $2 \cdot 5 \cdot 3 \cdot 7$

2. $5 \cdot 7 \cdot 2 \cdot 3$

Solutions

1. $2 \cdot 5 \cdot 3 \cdot 7$

$2 \cdot 5 \cdot 3 \cdot 7 = 10 \cdot 3 \cdot 7$

$10 \cdot 3 \cdot 7 = 30 \cdot 7$

$30 \cdot 7 = 210$

2. $5 \cdot 7 \cdot 2 \cdot 3$

$5 \cdot 7 \cdot 2 \cdot 3 = 35 \cdot 2 \cdot 3$

$35 \cdot 2 \cdot 3 = 70 \cdot 3$

$70 \cdot 3 = 210$

When skip counting by 5, the first number that you say is 5. This is the product of 1 and 5. The same is true when skip counting by any other whole number. This illustrates another important property, the **identity property of multiplication**.

Identity Property of Multiplication

When a is any number, then
$$1 \cdot a = a \quad \text{and} \quad a \cdot 1 = a$$
The number 1 is called the **multiplicative identity**.

Here are a couple examples of the identity property of multiplication:

$$1 \cdot 24 = 24 \cdot 1 = 24$$

$$1 \cdot 77 = 77 \cdot 1 = 77$$

Practice A

Multiply the same four numbers, 2, 3, 5, and 7, together in a different order than you saw in Example 1.

B. Exponent Notation

The layout of William's dishwasher rack from our previous example is more efficient than a standard rack. By using a honeycomb pattern, the rack is able to accommodate 6 rows of 5 tumbler slots while a standard rack can only accommodate 5 rows of 5 tumbler slots. How many tumblers can be fitted into 5 standard racks? We can represent this counting question with the product $5 \cdot 5 \cdot 5$.

In the same way that multiplication gives us a convenient way to represent repeated addition, exponent notation gives us a convenient way to represent repeated multiplication:

$$b^n = \underbrace{b \cdot b \cdot \ldots \cdot b}_{\text{There are } n \text{ factors.}}$$

The little superscript n above the b is called the **exponent**. It communicates the number of factors in the repeated product. Each factor has the same value, b, which is the **base**. We read b^n as "b raised to the nth power" or "b to the nth power."

If the exponent is either 0 or 1, it makes more sense to include a factor of 1 in front of the repeated product. We can do so without changing the result because of the identity property of multiplication. Here is the product that we end up with:

$$b^n = \underbrace{1 \cdot b \cdot b \cdot \ldots \cdot b}_{\text{There are } n + 1 \text{ factors.}}$$

Now n represents the number of multiplication symbols, and the number of factors is $n + 1$. The first factor is 1, and all of the other factors are b. We can now accurately interpret b^1 and b^0:

$b^1 = 1 \cdot b = b$: One multiplication symbol, the first factor is 1, and the remaining factor is b.

$b^0 = 1$: No multiplication symbols, the first factor is 1, and there are no other factors.

Because the area of a square with sides of length b is $b \cdot b = b^2$, we read b^2 as "b squared" instead of "b to the second power." Similarly, a cube shaped box with sides that are b by b squares has a volume of $b \cdot b \cdot b = b^3$, so we usually read b^3 as "b cubed" instead of "b to the third power."

Returning to William and the standard dishwasher racks, we now know that 5^3 tumblers can fit into 5 standard racks. To evaluate 5^3, we calculate $5^3 = 5 \cdot 5 \cdot 5 = 25 \cdot 5 = 125$, or we use the exponent feature on our calculator to directly calculate $5^3 = 125$.

Practice B

Evaluate the following expressions. Then turn the page to check your solutions.

1. 3^3 3. 6^2 5. 17^1

2. 2^4 4. 9^0

C. Division

We opened this section by describing the problem that William is trying to solve in his little mental math world of dishwashing. He wants to figure out how many dishwasher racks he'll need in order to accommodate all of the drinking glasses he has to wash.

The first step is to determine how many glasses fit on one rack. We used multiplication to answer this question, and we found that each rack accommodates 30 glasses.

Now suppose that lunch is ending at William's cafeteria, and there are now 240 dirty glasses that William needs to wash. If we divide the glasses into groups of 30, how many groups will there be? Each group of 30 glasses fits onto one rack, so the number of groups is the same as the number of racks that William needs. The answer to this question is called "the *quotient* of 240 and 30." We can also say that the answer is "240 *divided by* 30."

We can use skip counting to answer this question, too, but now we skip count by 30s until we reach 240, keeping track how many numbers we have named.

When we skip count by 30s, 240 is the eighth number that we name, so that means that the quotient of 240 and 30 is equal to 8. William will need 8 racks. This skip counting also represents the multiplication $8 \cdot 30 = 240$. This is the basis for the definition of division.

Practice A — Answers

No matter what order you use for multiplying, the associative and commutative properties of multiplication guarantee that you end up with the same product. In this case, that product is 210.

Division

Let a and b be any two numbers. If there is a unique number c such that $c \cdot b = a$, then we say that "a divided by b is equal to c." We can write this statement in the following ways:

$$a \div b = c \qquad \frac{a}{b} = c \qquad {}^{a}\!/_{b} = c \qquad b\overline{)a} = c$$

The number a is called the **dividend**, b is called the **divisor**, and c, the result of the division, is called the **quotient** of a and b.

Even though there are several different symbols used for division, they each mean "a divided by b is equal to c." We will see each of these symbols used throughout this book.

In the context of whole numbers, we often end up with divisions that are not addressed by this definition. For example, $20 \div 8$ appears to be undefined since there is no whole number c for which $c \cdot 8 = 20$. We will develop a more satisfactory solution to this problem later in the book, but for now, we'll simply say that the quotient of 20 and 8 is not a whole number.

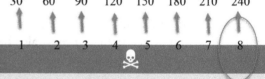

Figure 3

Warning – Incorrect Approach

The operation of multiplication complies with the commutative and associative properties. However, neither of these properties apply to division, so when we evaluate a sequence of divisions or a mixed sequence of divisions and multiplications, we must evaluate the operations in order from left to right.

$12 \div 3$ does not give the same result as $3 \div 12$

$(100 \div 2) \div 10$ does not give the same result as $100 \div (10 \div 2)$

Example 2

Evaluate the mixed sequence of divisions and multiplications $15 \cdot 20 \div 50 \cdot 7 \div 21$.

Solution

$15 \cdot 20 \div 50 \cdot 7 \div 21 = 300 \div 50 \cdot 7 \div 21$

$300 \div 50 \cdot 7 \div 21 = 6 \cdot 7 \div 21$

$6 \cdot 7 \div 21 = 42 \div 21$

$42 \div 21 = 2$

Practice C

Evaluate the following mixed sequence of divisions and multiplications. Then turn the page to check your solutions.

6. $24 \div 8 \cdot 30 \div 9$ 7. $12 \div 3 \cdot 2$ 8. $160 \div 5 \cdot 2 \div 16 \cdot 2$

D. Multiplication and Division Involving Zero

Consider the following four products and quotients. Let's use the definitions of multiplication and division to interpret and evaluate them.

$$5 \cdot 0 \qquad 0 \cdot 5 \qquad \frac{0}{5} \qquad \frac{5}{0}$$

The first expression is fairly straightforward. Five times zero means five zeros:

$$5 \cdot 0 = 0 + 0 + 0 + 0 + 0 = 0$$

The second one is a bit strange. Zero times five means zero fives. How do you even write that out in math symbols? How do you write zero *anything*? Fortunately, we know that multiplication is commutative, so if $5 \cdot 0 = 0$, then it must also be true that $0 \cdot 5 = 0$. This illustrates a fourth property of multiplication, the **zero product property**.

Zero Product Property of Multiplication

When a is any number, then

$$a \cdot 0 = 0 \quad \text{and} \quad 0 \cdot a = 0$$

Furthermore, if $a \cdot b = 0$, then at least one of the factors, a or b, must be 0.

Now consider the divisions that involve zero. First, is there a unique number c such that $c \cdot 5 = 0$? If there is such a number, then it is the quotient of 0 and 5. The zero product property tells us that there *is* such a number: $c \cdot 5 = 0$ if and only if $c = 0$. So, to answer our third problem, $\frac{0}{5} = 0$.

Finally, is there a unique number c such that $5 = 0 \cdot c$? If there is such a number, then it is the quotient of 5 and 0. But the zero product property tells us that no such number exists: $0 \cdot c$ must be

Practice B — Answers

1. $3^3 = 3 \cdot 3 \cdot 3$ 3. $6^2 = 6 \cdot 6$ 5. $17^1 = 17$
         $= 27$ $= 36$ $= 17$

2. $2^4 = 2 \cdot 2 \cdot 2 \cdot 2$ 4. $9^0 =$
         $= 16$ $= 1$

equal to zero and therefore cannot be equal to five. So $\frac{5}{0}$ is undefined. We now know two important facts about division. The first is the **zero dividend principle**.

Zero Dividend Principle

When a is any number *except* zero, then $\frac{0}{a} = 0$. Furthermore, if the quotient of two numbers is equal to zero, then the dividend *must be* zero.

In the same way that $\frac{5}{0}$ is undefined, every other number divided by 0 is also undefined. This is called the **zero divisor principle**.

Zero Divisor Principle

When a is any number *including* zero, $\frac{a}{0}$ is undefined.

The zero divisor principle applies even when the dividend is zero. $\frac{0}{0}$ is undefined. To see why this is the case, we turn again to the definition of division. Is there a *unique* number c such that $c \cdot 0 = 0$. If there is such a number, then it is the quotient of 0 and 0. In this case, there are many numbers that would work for c. In fact, $c \cdot 0 = 0$ no matter what number c represents. The quotient of 0 and 0 is undefined because the number c such that $c \cdot 0 = 0$ is not unique.

☠ Warning – Incorrect Approach ☠

Be careful when working with division problems that divide by zero! Because 0 is the additive identity, you might be tricked into writing $\frac{5}{0} = 5$, but that is never true! Any number divided by 0 is always undefined.

Zero can be a tricky, so let's try out a few practice problems.

Practice D

Evaluate the following products or quotients. Then turn the page to check your solutions.

9. $\frac{11}{0}$

10. $6 \cdot 0$

11. $\frac{0}{100}$

12. $0 \cdot 0$

13. $0 \cdot 9999$

14. $\frac{0}{0}$

Exercises 1.3

For the following exercises, evaluate each product.

1. $4 \cdot 7$
2. $3 \cdot 4$
3. $2 \cdot (4 \cdot 7)$
4. $(3 \cdot 4) \cdot 5$

5. $2 \cdot 4$
6. $4 \cdot 5$
7. $(2 \cdot 4) \cdot 7$
8. $3 \cdot (4 \cdot 5)$

9. $0 \cdot 38$
10. $87 \cdot 729 \cdot 58 \cdot 0$ (Don't use your calculator. Just think about it for a second.)

For the following exercises, write each product in expanded notation — in other words, as a repeated *sum*.

11. $6 \cdot 11$
12. $4 \cdot 2$
13. $4x$
14. $7 \cdot 3$

15. $2(6)$
16. $5a$
17. $1 \cdot 9$
18. $(8)3$

19. $9 \cdot 1$
20. $(3)(9)$

For the following exercises, write each power in expanded notation — in other words, as a repeated *product*.

21. 2^5
22. 4^2
23. 11^6
24. x^4

25. c^3
26. 23^5
27. 1^9
28. 1^2

29. 9^1
30. 2^1

For the following exercises, evaluate each quotient. Write "not a whole number" or "undefined" when appropriate.

31. $27 \div 9$
32. $35 \div 5$
33. $\frac{38}{11}$
34. $78 / 6$

35. $0 \div 37$
36. $\frac{6}{78}$
37. $20 \div 0$
38. $\frac{0}{16}$

39. $126 / 18$
40. $\frac{29}{0}$

Practice C — Answers

6. $24 \div 8 \cdot 30 \div 9 = 3 \cdot 30 \div 9$
    $= 90 \div 9$
    $= 10$

8. $160 \div 5 \cdot \div 16 \cdot 2 = 32 \cdot 2 \div 16 \cdot 2$
    $= 64 \div 16 \cdot 2$
    $= 4 \cdot 2$
    $= 8$

7. $12 \div 3 \cdot 2 = 4 \cdot 2$
    $= 8$

For the following exercises, pause to reflect. Then answer each question carefully.

41. Looking at your answers to questions 27 and 29, do you think that a^b and b^a have the same value when a and b are different numbers? Can you think of a case where they *do* have the same value and another case where they *do not* have the same value?

42. Looking at your answer to question 27, what do you think the value of 1^a is? Does it matter what number a represents?

43. Write out 0^3 in expanded form. What do you think the value of 0^a is? Does it matter what number a represents?

44. We learned in section 1.3 that $b^0 = 1$ when $b \neq 0$ (strange, but true), and in problem 43, you deduced that $0^a = 0$. Which of these two rules should you use to evaluate 0^0? This is an even-numbered problem, so the answer is not in the back of the book. If it bugs you, take this opportunity to find your professor's office!

45. Use the definition of division to evaluate $\frac{x}{x}$ without knowing anything about the number x except that x is not equal to zero.

For the following exercises, evaluate the mixed sequence of products and quotients.

46. $35 \div 7 \cdot 5$

47. $24 \div 4 \cdot 2$

48. $42 \div 3 \cdot 7$

49. $72 \div 3 \cdot 4 \div 2$

50. $56 \div 7 \cdot 8 \div 4$

51. $8 \cdot 18 \div 6 \div 3$

52. $2 \cdot 24 \div 3 \cdot 2$

53. $48 \div 6 \cdot 4 \div 2$

54. $27 \cdot 6 \div 2 \cdot 3$

For these final problems, we first return our attention to William the dishwasher and then look at two problems related to powers of 10.

55. When he takes the clean tumblers out of the dishwasher, William puts them away on trays. Each tray fits 4 rows of 7 tumblers. How many trays will William need to put away all of the 240 clean tumblers? How many tumblers will be on the last tray?

56. William decides to store the tumblers on a shelf that is two feet deep by four feet wide. If 36 tumblers fit on one square foot of the shelf, will the 240 tumblers all fit on the shelf? If not, how many will be left over?

57. Evaluate each power of 10. See if you can do so without a calculator.

 a. 10^2 b. 10^3 c. 10^4

58. If n is a whole number, propose a simple rule to evaluate 10^n by hand without needing to know any multiplication facts.

Practice D — Answers

9. $\frac{11}{0}$ is undefined because of the *zero divisor principle.*

10. $6 \cdot 0 = 0$ because of the *zero product property.*

11. $\frac{0}{100} = 0$ because of the *zero dividend property.*

12. $0 \cdot 0 =$ because of the *zero product property.*

13. $0 \cdot 9999 =$ because of the *zero product property.*

14. $\frac{0}{0}$ is undefined because of the *zero divisor property.*

1.4 Order of Operations

While studying for a big test one night, Enrique and Paul both attempt to evaluate $3 + 7 \cdot 2$. Paul calculates $3 + 7 \cdot 2 = 10 \cdot 2$ and comes up with 20 as the answer. Enrique calculates $3 + 7 \cdot 2 = 3 + 14$ and comes up with 17 as the answer. Neither evaluated a sum or a product incorrectly, but they came up with different answers because they evaluated the operations in different orders. However, they can't both be correct. Because the expression has no variable, it must have just one correct value.

In this section, we explore a set of rules to guide us through the various operations involved in any expression and make sure they happen in the right order. By the end of this section, we will use these rules to evaluate this expression:

$$7 \cdot 3^2 + \frac{(8 - 3) \cdot 6}{3 \cdot 5}$$

This complicated expression involves all of the rules for the order of operations.

A. First: Parentheses and Other Grouping Symbols

A grouping symbol is a symbol that separates one part of an expression from the rest of the expression. This is the first priority for evaluating an expression. Any grouped operations in the expression must be evaluated before any other operations can be evaluated.

Here is a partial list of grouping symbols:

♦ parentheses of various types, including (parentheses), [brackets], and {braces}

♦ exponents

♦ the fraction bar division symbol

There are other grouping symbols that we will learn about later in this book. Here are some examples to illustrate how these grouping symbols work.

> **Example 1**
>
> Evaluate the following expressions.
>
> **1.** $[4 \cdot (5 - 3)]^2$ **2.** 2^{3+5} **3.** $\frac{9 + 7}{2^3}$
>
> **Solutions**
>
> **1.** $[4 \cdot (5 - 3)]^2$
>
> This is a power expression. The exponent is 2, and the base is the number value of the part of the expression that is grouped by the brackets: $4 \cdot (5 - 3)$. But this grouped part includes a sub-grouped operation within parenthesis: $5 - 3$. To simplify this expression, we start from the inner-most grouping and work our way to the outermost grouping.

$$[4 \cdot (5 - 3)]^2 = [4 \cdot 2]^2$$

$$[4 \cdot 2]^2 = 8^2$$

$$= 64$$

Notice that we remove the parentheses after a grouped or sub-grouped part has been evaluated.

2. 2^{3+5}

This is a power expression. In this expression, the base is 2, and the exponent is the number value of the part that is grouped by the superscript: $3 + 5$. In this expression, the addition is *grouped* because it is in the exponent. To evaluate this expression, we must first evaluate the grouped part so that we know what the exponent is.

$$2^{3+5} = 2^8$$

$$= 256$$

3. $\dfrac{9 + 7}{2^3}$

This is a quotient. The fraction bar is a division symbol, but it also serves as a grouping symbol. The dividend is the number value of the grouped part *above* the fraction bar, $9 + 7$. The divisor is the number value of the grouped part *below* the fraction bar, 2^3. To evaluate this expression, we must first evaluate these grouped parts so that we know the correct values for the dividend and divisor of the quotient.

$$\frac{9 + 7}{2^3} = \frac{16}{8}$$

$$= 2$$

Let's turn to our running example for this section and identify any grouped and sub-grouped parts:

$$7 \cdot 3^2 + \frac{(8 - 3) \cdot 6}{3 \cdot 5}$$

The parts of the expression after the plus sign are grouped by the fraction bar. The grouped part above the fraction bar, $(8 - 3) \cdot 6$, includes a sub-grouped subtraction in parentheses. Here is how we evaluate the grouped parts of this expression:

$$7 \cdot 3^2 + \frac{(8 - 3) \cdot 6}{3 \cdot 5} = 7 \cdot 3^2 + \frac{5 \cdot 6}{3 \cdot 5}$$

$$= 7 \cdot 3^2 + \frac{30}{15}$$

Now the expression no longer has any grouped parts.

Practice A

Evaluate the following expressions. Then turn the page and check your answers.

1. $[(3 - 2) + 1]^{1+1}$

2. $\dfrac{3^2}{1 + 2}$

3. $\left(\dfrac{7 + 1}{2^{3-1}}\right)^{2-1}$

4. $\dfrac{2^{3 \cdot 2}}{4 \cdot 4}$

5. $[2 + (7 - 3)]^{7-5}$

6. $\dfrac{2 - (5 - 3)}{4 \cdot (6 - 1)}$

B. Second: Exponents

After all of the grouped parts of the expression have been evaluated, the second priority is the exponents that remain. Ungrouped powers must be evaluated before any other ungrouped operations can be evaluated.

Consider how the expressions 2^{3+5} and $2^3 + 5$ are different. In the first expression, the addition is grouped, so it must be evaluated before the power is evaluated:

$$2^{3+5} = 2^8$$

$$= 256$$

In the second expression, the addition is not grouped, so the power must be evaluated first:

$$2^3 + 5 = 8 + 5$$

$$= 13$$

In our running example, all of the grouped parts have been evaluated, so we are left with $7 \cdot 3^2 + \dfrac{30}{15}$. Now we must evaluate the power. The exponent is 2, but what is the base? Is the base 3 or 21? Because we must evaluate the power *before* the multiplication, the base cannot be 21.

$$7 \cdot 3^2 + \dfrac{30}{15} = 7 \cdot 9 + \dfrac{30}{15}$$

Practice B

Evaluate the following expressions. Then turn the page and check your solutions.

7. $4^{1+1} + 2$

8. $5^2 \cdot 3$

9. $12 - (5 - 2)^2$

10. $2 \cdot 4^3$

11. $24 \div 2^{4-1}$

12. $5 + \left[\dfrac{(6 - 2)^2}{3 + 1}\right]^{2+1}$

C. Third: Multiplication and Division

After all of the grouped parts of an expression and all of the ungrouped powers have been evaluated, our third priority is to evaluate any products and quotients that remain. Ungrouped products and quotients must be evaluated before any ungrouped additions or subtractions are evaluated. And remember — products and quotients are evaluated in order *from left to right.*

Let's return to the example of Enrique and Paul. They both attempt to evaluate $3 + 7 \cdot 2$. Paul calculates $3 + 7 \cdot 2 = 10 \cdot 2$ and comes up with 20 as the answer. Enrique calculates $3 + 7 \cdot 2 = 3 + 14$ and comes up with 17 as the answer. This expression has no grouped parts and no exponents, so we have to evaluate all products and quotients first. This is what Enrique did. He evaluated the multiplication of $7 \cdot 2$ before adding that to 3. Paul's answer is incorrect because he added 3 and 7 before evaluating the multiplication with 2. He did not follow this third priority.

With the complicated example from the start of this section, we are now reduced to a multiplication, an addition, and a division. This third rule tells us the multiplication and division must be evaluated in the order that they appear from left to right, and both must be evaluated before the addition can be evaluated.

$$7 \cdot 9 + \frac{30}{15} = 63 + 2$$

Practice C

State whether or not each of the following equations are true. Explain why or why not. Then turn the page and check your solutions.

13. $3 \cdot 2 + 6 = 24$

14. $\frac{44}{11} + 2 \cdot 2 = 8$

15. $7 \cdot 13 - 12 = 7$

16. $1 + 2 \cdot (3 + 3) = 18$

17. $12 \div 2 \cdot 3 - 18 \div 6 \cdot 3 = 9$

18. $2 + 3 \cdot 4 = 14$

D. Fourth: Addition and Subtraction

After all the grouped parts, exponents, and products and quotients have been evaluated, the only operations remaining should be ungrouped sums and differences. These should be evaluated in order, moving from left to right.

After completing the first, second, and third priorities, the complicated running example from the beginning of this section has been reduced to $63 + 2$, so the final value of the expression is 65. Here is the complete process that we used to evaluate this expression:

$$7 \cdot 3^2 + \frac{(8 - 3) \cdot 6}{3 \cdot 5} = 7 \cdot 3^2 + \frac{30}{15} \qquad \text{First priority: Parentheses}$$

$$= 7 \cdot 9 + \frac{30}{15} \qquad \text{Second priority: Exponents}$$

$$= 63 + 2 \qquad \text{Third priority: Multiplications and Divisions}$$

$$= 65 \qquad \text{Fourth priority: Additions and Subtractions}$$

The colored initial letters spell out the nonsense word PEMDAS, which many students use to help remember the correct order of operations. Some students remember PEMDAS as an acronym for "Please Excuse My Dear Aunt Sally." You can make up your own phrase if you like, such as "Purple Elephants Might Drink Avocado Soda," or something even better.

Please keep in mind that although there are six letters in this word, there are only four priorities in the order of operations. Multiplication and division have the same priority as each other. Addition and subtraction are also the same priority as each other.

Practice A — Answers

1. $[(3 - 2) + 1]^{1+1} = [1 + 1]^2$
    $= [2]^2$
    $= 4$

2. $\dfrac{3^2}{1 + 2} = \dfrac{3^2}{3}$
    $= \dfrac{9}{3}$
    $= 3$

3. $\left(\dfrac{7 + 1}{2^{3-1}}\right)^{2-1} = \left(\dfrac{8}{2^2}\right)^1$
    $= \left(\dfrac{8}{4}\right)^1$
    $= 2^1$
    $= 2$

4. $\dfrac{2^{3+2}}{4 \cdot 4} = \dfrac{2^6}{16}$
    $= \dfrac{64}{16}$
    $= 4$

5. $[2 + (7 - 3)]^{7-5} = [2 + 4]^2$
    $= 6^2$
    $= 36$

6. $\dfrac{2 - (5 - 3)}{4 \cdot (6 - 1)} = \dfrac{2 - 2}{4 \cdot 5}$
    $= \dfrac{0}{20}$
    $= 0$

Practice B — Answers

7. $4^{1+1} + 2 = 4^2 + 2$
    $= 16 + 2$
    $= 18$

8. $5^2 \cdot 3 = 25 \cdot 3$
    $= 75$

9. $12 - (5 - 2)^2 = 12 - 3^2$
    $= 12 - 9$
    $= 3$

10. $2 \cdot 4^3 = 2 \cdot 64$
    $= 128$

11. $24 \div 2^{4-1} = 24 \div 2^3$
    $= 24 \div 8$
    $= 3$

12. $5 + \left[\dfrac{(6 - 2)^2}{3 + 1}\right]^{2+1} = 5 + \left[\dfrac{4^2}{4}\right]^{2+1}$
    $= 5 + \left[\dfrac{16}{4}\right]^{2+1}$
    $= 5 + 4^{2+1}$
    $= 5 + 4^3$
    $= 5 + 64$
    $= 69$

Practice D

Evaluate the following expressions using the correct order of operations. Then turn the page and check your solutions.

19. $12 - (3 + 4) + 1$

20. $2 \cdot 3 \cdot 4 - 2 \cdot 3 + 4$

21. $2 \cdot 3^4 \div 9 - 2 + 1$

22. $4^{5-2} - 2^4 - 9 + 4 \cdot 2$

23. $24 \div 2 \cdot 3 - 4 \cdot (4 - 1) + 3^2$

24. $25 - 2 \cdot \left[\frac{(5-3)^2}{2} \right]^{2+1} + 7$

E. Evaluating Grouped Parts

When evaluating the grouped parts of an expression, all four of these priority levels must be observed within each grouped part.

Example 2

Evaluate the expression: $\dfrac{6 \cdot (14 - 3^2)^2}{15}$

Solution

The first step in evaluating this expression is to evaluate the part of the expression that is grouped above the fraction bar: $6 \cdot (14 - 3^2)^2$. But to simplify this expression, we must observe the correct order of operations:

$6 \cdot (14 - 3^2)^2 \ = \ 6 \cdot (14 - 9)^2$ First simplify the grouped part of the expression above the fraction bar. In this grouped part, the exponent must be evaluated before the subtraction.

Practice C — Answers

13. No, $3 \cdot 2 + 6 = 24$ is not true. Multiplication needs to be performed before addition, so $3 \cdot 2 + 6$ simplifies to 12 and not 24

14. Yes, $\frac{44}{11} + 2 \cdot 2 = 8$ is true. The quotient and the product must be evaluated first, resulting in the equation $4 + 4 = 8$. Then we evaluate the sum to obtain the true statement $8 = 8$.

15. No, $7 \cdot 13 - 12 = 7$ is not true. Multiplication needs to be performed before subtraction, so $7 \cdot 13 - 12$ simplifies to 79 and not 7.

16. No, $1 + 2 \cdot (3 + 3) = 18$ is not true. First simplify $3 + 3$ in the parentheses, and then perform the multiplication before the remaining addition, so $1 + 2 \cdot (3 + 3)$ simplifies to 13 and not 18.

17. Yes, $12 \div 2 \cdot 3 - 18 \div 6 \cdot 3 = 9$ is true. Since multiplication and division are performed before subtraction, first perform the multiplication and divisions from left to right, so the equation $12 \div 2 \cdot 3 - 18 \div 6 \cdot 3 = 9$ becomes $18 - 9 = 9$. Then we evaluate the difference to obtain the true statement $9 = 9$.

18. Yes, $2 + 3 \cdot 4 = 14$ is true. Multiplication needs to be performed before addition, so $2 + 3 \cdot 4$ simplifies to 14. the resulting equation, $14 = 14$, is a true statement.

$$6 \cdot (14 - 9)^2 = 6 \cdot 5^2 \qquad \text{Now evaluate the subtraction within the grouping and remove the parentheses.}$$

$$6 \cdot 5^2 = 6 \cdot 25 \qquad \text{With the grouped part of this expression gone, evaluate the exponents.}$$

$$6 \cdot 25 = 150 \qquad \text{Last comes multiplication and division.}$$

A grouped part of an expression should be evaluated as if it were an expression in its own right, independent of the larger expression in which it is embedded. Now we finish by simplifying the expression.

$$\frac{6 \cdot (14 - 3^2)^2}{15} = \frac{150}{15}$$
$$= 10$$

☠ Warning – Incorrect Approach ☠

Some calculators are not able to follow the correct order of operations when solving a problem. You may have seen some social media posts that have problems meant to trip up a calculator. The arguments that ensue are unnecessary and easily resolved by knowing the correct order of operations. Gain a thorough understanding of both the order of operations as well as their priorities instead of relying on your calculator to always provide the right answer.

Practice E

Evaluate the following expressions. Then turn the page and check your solutions.

25. $7^{4 + 3 - 10}$

26. $(4 + 3 \cdot 2)^2 + 10$

27. $\dfrac{12 \cdot (34 - 2^5) - 3^2}{3 \cdot 4 - 7}$

28. $\dfrac{2 + 3 \cdot 4 - 2^2}{2 \cdot 3 + 4}$

29. $(24 \div 6 \cdot 2)^2$

30. $(8 - 2 - 3 + 1)^3 - 2^{3 \cdot 2 - 1}$

Exercises 1.4

For the following exercises, write the *first* operation that should be evaluated in each expression.

1. $2 + 3(4)$

2. $4 + 3^2$

3. $85 - 12 \cdot 6$

4. $40 \div 10 \cdot 2$

5. $(4 \cdot 3)^2$

6. $2^{3 + 4}$

7. $18 - 12 + 5$

8. $5^{4 - 4}$

9. $(2^3)^4$

10. $\left(\dfrac{4}{2}\right)^3$

11. $5^4 - 4$

12. $\dfrac{3 \cdot (8 + 2 \cdot 3)}{21}$

13. $\dfrac{3}{5 - 5}$

14. $\dfrac{7}{3^2 - 9}$

15. $24 \div 6 \cdot 2$

16. $2 + 3(5)$

17. $2^{3(4)}$

18. $24 - 3 + 1 + 2^3 + 5$

19. $\dfrac{90}{(5 + 5 \cdot 2)3}$

20. $17 - 4 + 2 + 1$

21. $48 - 3 + 2 \cdot 3 - 1$

22. $\dfrac{24}{3 - 2 + 1}$

23. $3^{4 - 3}$

24. $4^{2 + 1} + 5$

For the following exercises, evaluate each expression. If the result is undefined, write "undefined."

25. $2 + 3(4)$

26. $4 + 3^2$

27. $85 - 12 \cdot 6$

28. $40 \div 10 \cdot 2$

29. $(4 \cdot 3)^2$

30. 2^{3+4}

31. $18 - 12 + 5$

32. 5^{4-4}

33. $(2^3)^4$

34. $\left(\frac{4}{2}\right)^3$

35. $5^4 - 4$

36. $\frac{3 \cdot (8 + 2 \cdot 3)}{21}$

37. $\frac{3}{5 - 5}$

38. $\frac{7}{3^2 - 9}$

39. $24 \div 6 \cdot 2$

40. $2 + 3(5)$

41. $2^{3(4)}$

42. $24 - 3 + 1 + 2^3 + 5$

43. $\frac{90}{(5 + 5 \cdot 2)3}$

44. $17 - 4 + 2 + 1$

Practice D — Answers

19.
$$12 - (3 + 4) + 1 = 12 - 7 + 1$$
$$= 5 + 1$$
$$= 6$$

20.
$$2 \cdot 3 \cdot 4 - 2 \cdot 3 + 4 = 6 \cdot 4 - 2 \cdot 3 + 4$$
$$= 24 - 2 \cdot 3 + 4$$
$$= 24 - 6 + 4$$
$$= 18 + 4$$
$$= 22$$

21.
$$2 \cdot 3^4 \div 9 - 2 + 1 = 2 \cdot 81 \div 9 - 2 + 1$$
$$= 162 \div 9 - 2 + 1$$
$$= 18 - 2 + 1$$
$$= 16 + 1$$
$$= 17$$

22.
$$4^{5-2} - 2^4 - 9 + 4 \cdot 2 = 4^3 - 2^4 - 9 + 4 \cdot 2$$
$$= 64 - 16 - 9 + 4 \cdot 2$$
$$= 64 - 16 - 9 + 8$$
$$= 48 - 9 + 8$$
$$= 39 + 8$$
$$= 47$$

23.
$$24 \div 2 \cdot 3 - 4 \cdot (4 - 1) + 3^2 = 24 \div 2 \cdot 3 - 4 \cdot 3 + 3^2$$
$$= 24 \div 2 \cdot 3 - 4 \cdot 3 + 9$$
$$= 12 \cdot 3 - 4 \cdot 3 + 9$$
$$= 36 - 4 \cdot 3 + 9$$
$$= 36 - 12 + 9$$
$$= 24 + 9$$
$$= 33$$

24.
$$25 - 2 \cdot \left[\frac{(5-3)^2}{2}\right]^{2+1} + 7 = 25 - 2 \cdot \left[\frac{2^2}{2}\right]^3 + 7$$
$$= 25 - 2 \cdot \left[\frac{4}{2}\right]^3 + 7$$
$$= 25 - 2 \cdot 2^3 + 7$$
$$= 25 - 2 \cdot 8 + 7$$
$$= 25 - 16 + 7$$
$$= 9 + 7$$
$$= 16$$

45. $48 - 3 + 2 \cdot 3 - 1$

46. $\dfrac{24}{3 - 2 + 1}$

47. 3^{4-3}

48. $7 \cdot 5 - 3 \cdot 8$

49. $\dfrac{(5 \cdot 3 - 4 \div 2)^2}{4 \cdot 3 - 24 \div 2}$

50. $\dfrac{24 \div 2^2 - 2 \cdot 3}{(6 \cdot 3 - 4 \div 2)^2}$

51. $\dfrac{(3 + 5)^2}{7 \cdot 3 - 5}$

52. $(3 \cdot 7 - 2 \cdot 4)^{2 \cdot 3 - (7-5)}$

53. $3 \cdot \left[4 \cdot 6 \div 3 + 3 - \dfrac{10}{10 - 4 - 4} \right]$

54. $\dfrac{24 \div 12 \cdot 2}{16 \div 2^3 \cdot 2}$

55. $\dfrac{[9 - 3^2 \cdot (6 - 2) \div 12]^2}{3 \cdot 4^3 \div 16}$

56. $\left[4 \cdot \left(\dfrac{10 + 4 \cdot 6}{1 + 4^2} \right)^3 - 7 \right] \div 5$

57. $(2 \cdot 3 \cdot 4 - 12 \div 2 \cdot 3)^{2 \cdot 3 - 4}$

58. $\dfrac{(5 - 2)^{11 - 3 \cdot 3}}{5 \cdot 3 - 12}$

For the following exercises, pause to reflect first. Then answer each question carefully.

59. Suppose you wish to use your calculator to evaluate the expression $\dfrac{7 + 8}{4 - 1}$. The only division symbol on your calculator's keypad is the inline division symbol ÷, so you enter the expression $7 + 8 \div 4 - 1$ into your calculator. Will you end up with the correct answer? If you answer no, then how should you have entered the expression into the calculator?

60. To evaluate 2^{3^2}, which way is correct: $2^{3^2} = 8^2 = 64$ or $2^{3^2} = 2^9 = 512$? Hint: is any part of this expression grouped? Review the sub-section about grouping symbols carefully.

Practice E — Answers

25.
$$7^{4 \cdot 3 - 10} = 7^{12 - 10}$$
$$= 7^2$$
$$= 49$$

26.
$$(4 + 3 \cdot 2)^2 + 10 = (4 + 6)^2 + 10$$
$$= 10^2 + 10$$
$$= 100 + 10$$
$$= 110$$

27.
$$\dfrac{12 \cdot (34 - 2^5) - 3^2}{3 \cdot 4 - 7} = \dfrac{12 \cdot (34 - 32) - 3^2}{12 - 7}$$
$$= \dfrac{12 \cdot 2 - 3^2}{5}$$
$$= \dfrac{12 \cdot 2 - 9}{5}$$
$$= \dfrac{24 - 9}{5}$$
$$= \dfrac{15}{5}$$
$$= 3$$

28.
$$\dfrac{2 + 3 \cdot 4 - 2^2}{2 \cdot 3 + 4} = \dfrac{2 + 3 \cdot 4 - 4}{6 + 4}$$
$$= \dfrac{2 + 12 - 4}{10}$$
$$= \dfrac{14 - 4}{10}$$
$$= \dfrac{10}{10}$$
$$= 1$$

29.
$$(24 \div 6 \cdot 2)^2 = (4 \cdot 2)^2$$
$$= 8^2$$
$$= 64$$

30.
$$(8 - 2 - 3 + 1)^3 - 2^{3 \cdot 2 - 1} = (6 - 3 + 1)^3 - 2^{6-1}$$
$$= (3 + 1)^3 - 2^5$$
$$= 4^3 - 2^5$$
$$= 64 - 32$$
$$= 32$$

1.5 Solving Equations

In Section 1.2, we learned how to use addition/subtraction fact families to solve certain types of equations. Addition/subtraction fact families arise from the definition of subtraction.

You may have noticed that our definition of division in Section 1.3 is remarkably similar to our definition of subtraction in Section 1.1. We use multiplication to define division in almost the same way that we use addition to define subtraction. In fact, the definition of division gives rise to multiplication/division fact families in the same way that addition/subtraction fact families arise from the definition of subtraction.

In this section, we learn how to use multiplication/division fact families to solve simple equations involving only one multiplication or one division, and we will also learn how to solve more complicated two-step equations that involve more than one operation. By the end of this section, you'll find that solving two-step equations is as simple as taking off your shoes and socks.

A. Multiplication/Division Fact Families

The definition of division demonstrates how every division fact corresponds to a multiplication fact. Every multiplication or division fact is part of a family of four equivalent equations called a **multiplication/division fact family**.

Multiplication/Division Fact Family

Assigning number values or variables to a, b, and c in the following four equations results in a multiplication/division fact family:

1. $\frac{a}{b} = c$

2. $c \cdot b = a$

3. $b \cdot c = a$

4. $\frac{a}{c} = b$

Equations 1 and 2 are equivalent to each other because of the *definition of division*.
Equations 2 and 3 are equivalent to each other because of the *commutative property of multiplication*.
Equations 3 and 4 are equivalent to each other because of the *definition of division*.

As with addition/subtraction fact families, we can use the symmetric property of equality to write an equation that is equivalent to any given equation by moving the expression on the left side of the equals sign to the right side and moving the expression on the right side to the left.

Example 1

Write out the complete multiplication/division fact family for $5x = 70$.

Solution

If we use the second equation in the fact family, $c \cdot b = a$, where $c = 5$, $b = x$. and $a = 70$, the equation numbers below correspond to the ones in the fact family definition box.

1. $\frac{70}{x} = 5$

2. $5x = 70$

3. $x5 = 70$

4. $\frac{70}{5} = x$

We can use multiplication/division fact families to solve equations in which the variable is a part of a product or quotient in the same way that we use addition/subtraction fact families to solve equations in which the variable is a part of a sum or difference.

Example 2

Solve the equation: $5x = 70$

Solution

First, we select the equation from the fact family for $5x = 70$ in which the variable is alone on one side of the equal sign. Then we solve that equation by evaluating the product or quotient.

$$5x = 70 \qquad \text{We start with the original equation.}$$
$$\frac{70}{5} = x \qquad \text{From the fact family, we choose equation 4 to solve.}$$
$$14 = x$$

Now we check our solution.

$$5 \cdot 14 = 70$$
$$70 = 70$$

Practice A

Write out the complete multiplication/division fact families for the equations. Next solve the equations. Then turn the page to check your solutions.

1. $\frac{72}{k} = 6$

2. $39 = 3m$

3. $11 = \frac{d}{7}$

Exercises 1.5

Write out the complete fact family for each equation, then identify any equations that have the variable alone on one side of the equation.

1. $5p = 40$

2. $5 + v = 40$

3. $\frac{65}{z} = 5$

4. $23 = \frac{b}{17}$

5. $m - 28 = 53$

6. $74 - k = 39$

7. $23 = 12 + b$

8. $24 = 6a$

9. $24 = \frac{u}{3}$

10. $143 = 670 - f$

11. $6u = 36$

12. $7w = 42$

13. $21 + b = 62$

14. $\frac{51}{g} = 17$

15. $20 = \frac{f}{7}$

16. $x - 49 = 67$

17. $93 = 3k$

18. $x + 142 = 345$

19. $80 = 512 - z$

20. $413 = t - 211$

21. $\frac{6}{110} = 6$

22. $2 = \frac{116}{e}$

23. $621 = 396 + w$

24. $475 = 25s$

For the following exercises, solve each equation. Be sure to check your solutions.

25. $5p = 40$

26. $5 + v = 40$

27. $\frac{65}{z} = 5$

28. $23 = \frac{b}{17}$

29. $m - 28 = 53$

30. $74 - k = 39$

31. $23 = 12 + b$

32. $24 = 6a$

33. $24 = \frac{u}{3}$

34. $143 = 670 - f$

35. $6u = 36$

36. $7w = 42$

37. $21 + b = 62$

38. $\frac{51}{g} = 17$

39. $20 = \frac{f}{7}$

40. $x - 49 + 67$

41. $93 = 3k$

42. $x + 142 = 345$

43. $80 = 512 - z$

44. $413 = t - 211$

45. $\frac{6}{110} = 6$

46. $2 = \frac{116}{e}$

47. $621 = 396 + w$

48. $475 = 25s$

49. $\frac{x}{5} = 230$

50. $\frac{116}{b} = 4$

Practice A – Answers

1. The equation $\frac{72}{k} = 6$ has the same format as equations 1 and 4 of the fact family definition box. If we use the first equation in the fact family, $\frac{a}{b} = c$, where $a = 72$, $b = k$, and $c = 6$, the equation numbers below correspond to the ones in the fact family definition box.

    a. $\frac{72}{k} = 6$ c. $k6 = 72$

    b. $6k = 72$ d. $k = \frac{72}{6}$

    And this is the solution:

    $$\frac{72}{k} = 6$$

    $$\frac{72}{6} = k$$

    $$12 = k$$

    And here is how we check the solution:

    $$\frac{72}{12} = 6$$

    $$6 = 6$$

2. The equation $39 = 3m$ has the same format as equations 2 and 3 of the fact family definition box. If we use the second equation in the fact family, $c \cdot b = a$, where $a = 39$, $b = m$, and $c = 3$, the equation numbers below correspond to the ones in the fact family definition box.

    a. $\frac{39}{m} = 3$ c. $m \cdot 3 = 39$

    b. $3m = 39$ d. $\frac{39}{3} = m$

    And this is the solution:

    $$39 = 3m$$

    $$\frac{39}{3} = m$$

    $$13 = m$$

    And here is how we check the solution:

    $$39 = 3 \cdot 13$$

    $$39 = 39$$

3. The equation $11 = \frac{d}{7}$ has the same format as equations 2 and 3 of the fact family definition box. If we use the second equation in the fact family, $c \cdot b = a$, where $a = d$, $b = 7$, and c = 11, the equation numbers below correspond to the ones in the fact family definition box.

a. $\frac{d}{7} = 11$

b. $11 \cdot 7 = d$

c. $7 \cdot 11 = d$

d. $\frac{d}{11} = 7$

And this is the solution:

$$11 = \frac{d}{7}$$
$$11 \cdot 7 = d$$
$$77 = d$$

And here is how we check the solution:

$$11 = \frac{77}{7}$$
$$11 = 11$$

CHAPTER 2
Integers

What Are Numbers?

The answer to this question in Chapter 1 was the set of whole numbers. We understood numbers to be symbols that are used to count collections of objects. However, some scenarios can't be fully addressed by the set of whole numbers.

Suppose Jose buys a house for $200,000. He can scrape together a down payment of $30,000, so he owes $170,000. Meanwhile, Shruti, who is exceptionally lucky, finds a winning lottery ticket on the sidewalk and turns it in for a prize of exactly $170,000. So Jose has $170,000 *less* than zero, while Shruti has $170,000 *more* than zero. We cannot say that they have the same amount of money.

Similarly, if we use whole numbers to represent temperature measurements, we would have to use the same number to describe two very different temperatures: 50° *above* zero, which means you can go outside to plant your garden, and 50° *below* zero, which means you are getting tired of winters in North Dakota. Does it seem reasonable to use the same number to describe both temperatures?

We can see how tricky these numbers are when we think about elevation. The deepest point on the ocean floor is the Challenger Deep at the bottom of Mariana Trench, where the elevation is 36,070 feet *below* sea level. In contrast, the highest point on Earth's surface is the peak of Mt. Everest, which is at an elevation of 29,029 feet *above* sea level. If we get into a submarine and descend to the bottom of the Challenger Deep, we would eventually be 29,029 feet *below* sea level, but we would certainly not be at the same elevation as the summit of Mt. Everest! The elevation beneath the sea is different than above the sea.

Is it reasonable to use the same number to describe these two very different types of numbers that are above or below zero? No, it isn't.

One limitation with the set of whole numbers is that there is no whole number that is less than zero. This not only restricts the usefulness of the set of whole numbers, it also introduces a lack of symmetry. Whole numbers are unbounded in one direction because there is no greatest whole number, but they are bounded in the other direction because 0 is the least of all the whole numbers.

In this chapter, we expand our definition of number to include numbers that are the opposite of whole numbers. This new set of numbers, the **integers**, is the set $\{\ldots, -4, -3, -2, -1, 0, 1, 2, 3, 4, \ldots\}$.

2.1 Introduction to Integers

Each positive integer corresponds with a negative integer. For example, the positive integer 7 corresponds to the negative integer –7. These corresponding integers are called **opposites** of each other. When the subtraction symbol is placed in front of a positive integer in such a way that it cannot possibly indicate the operation of subtraction, it indicates that the integer is a negative integer.

A. The Number Line

Imagine a straight line that extends infinitely far in both directions. Evenly spaced ticks are marked out all along the line. Each tick corresponds to an integer. All of the positive integers extend from the zero tick to the right in increasing order. All of the negative integers extend from the zero tick to the left in decreasing order. This line is called the **number line**. It is a visual aid for understanding the set of integers:

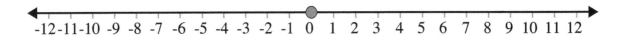

Greater Than and Less Than

Let a and b represent any two integers. If a is to the left of b on the number line then we say "a is less than b." Using math symbols, we write $a < b$.

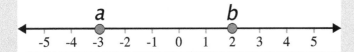

If a is to the right of b then we say "a is greater than b." Using math symbols, we write $a > b$.

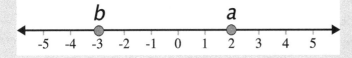

Example 1

Use "is greater than" or "is less than" to complete the following comparisons. Then write the comparisons using symbols.

1. 7__3   **2.** –7__3

Solutions

1. Because 7 is positioned to the right of 3 on the number line we say 7 *is greater than* 3, which we write symbolically as 7 > 3. This fits naturally with our understanding of whole numbers. If you have 7 bananas, then you have more than 3 bananas.

2. Because –7 is positioned to the left of 3 on the number line, we say –7 *is less than* 3, which we write symbolically as –7 < 3. It is tough to fit this into a banana counting situation, but it makes sense if we think of temperatures. Seven degrees below zero is colder than 3 degree above zero. Every negative integer is automatically less than any positive integer.

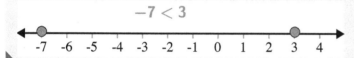

Practice A

Use positions on the number line to fill in each blank with the correct symbol, either < (is less than) or > (is greater than). Then check your answers at the end of this section.

1. 5__–3   4. 3__4

2. –9__–2   5. –3__–4

3. –10__8   6. –9__–7

B. Absolute Value

Every integer is represented by a tick mark on the number line. We measure distance on the number line by using the gap between two consecutive tick marks as one unit. For example, we say that the number line distance between 8 and 5 is 3 units because there are 3 gaps between 8 and 5 on the number line.

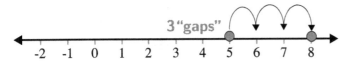

Keep in mind that the number line distance between two integers will always be a whole number since it is the result of a counting problem. We use this measurement of number line distance to make the following definition of **absolute value**.

Absolute Value

Let *a* represent any integer. The **absolute value** of *a* is the number line distance between zero and *a*. The math symbol for "absolute value of *a*" is $|a|$.

An absolute value can never be negative in the same way distance can never be negative. Imagine if your doctor ordered you to start running negative 2 miles every day — how would you even begin?

Example 2

Evaluate the following absolute values.

   1. $|-6|$ 2. $|6|$

Solutions

1. $|-6| = 6$ because the number line distance between −6 and 0 is 6.

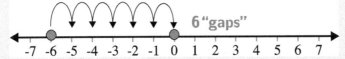

2. $|6| = 6$ because the number line distance between 6 and 0 is 6.

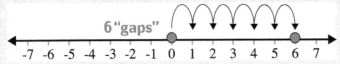

Notice from this example that integers that are opposites of each other always have the same absolute value. Each integer is the same distance from zero as its opposite — but in the opposite direction.

Practice B

Evaluate the following absolute values, then check your answers at the end of this section.

7. $|4|$ 9. $|1|$ 11. $|-1|$

8. $|-9|$ 10. $|-2|$ 12. $|3|$

C. Opposite vs. Negative

In the introduction to this section, we saw that when the minus sign is placed in front of a positive integer in such a way that it cannot possibly indicate the operation of subtraction, then it indicates a negative integer. By this rule, the minus sign (–) means subtraction in 9 – 4, but it means negative in –4 + 9 because it can't mean subtraction if there is no number before it. The minus sign also means negative in 9 + –4 because it cannot mean subtraction. It doesn't make sense to add and subtract at the same time. In this situation, we usually put the negative number in parentheses just to make the meaning clearer: 9 + (–4).

Sometimes we place the "–" symbol in front of a negative integer or a variable that might represent a negative integer in a way that can't possibly indicate the operation of subtraction. In this scenario, the "–" symbol means opposite. For example, – (–12) means *the opposite of* –12. Because –12 and 12 are opposites of each other, the opposite of –12 must be 12. In symbols, we write – (–12)= 12.

If the variable x could represent any integer, then $-x$ means "the *opposite* of x" and not necessarily "negative x" because we don't know whether x is a positive or negative integer.

Practice A — Answers

1. 5 > –3

4. 3 < 4

2. –9 < –2

5. –3 > –4

3. –10 < 8

6. –9 < –7

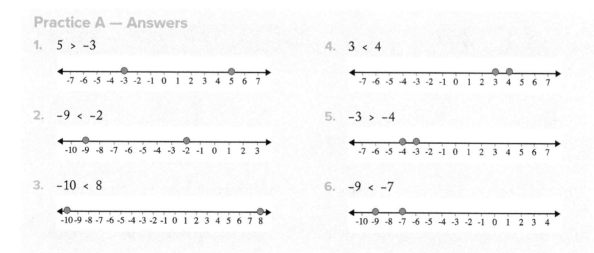

Example 3

1. Evaluate $-x$ when $x = 2$

2. Evaluate $-x$ when $x = -4$

3. Evaluate $-x$ when $x = |-7|$

Solution

1. To evaluate $-x$, we replace the variable symbol x with the given integer value 2, which we will put in parentheses to make the meaning clearer: $-x = -(2) = -2$. Notice that $-x$ turns out to be a negative integer when x is a positive integer.

2. To evaluate $-x$, we replace the variable symbol x with the given integer value -4, which we will put in parentheses to make the meaning clearer: $-x = -(-4) = 4$. Notice that $-x$ turns out to be positive integer when x is a negative integer.

3. To evaluate $-x$, we replace the variable symbol x with $|-7|$, which results in the expression $-|-7|$. In words, we want to find the opposite of the absolute value of negative 7. The absolute value of -7 is 7, so $-|-7| = -7$. Tricky!

Practice C

Read the following expressions out loud and be careful to use the appropriate words. Then evaluate the expression. Check your answers at the end of this section.

13. $-(-5)$

14. $-|-5|$

15. $-x$ when $x = 23$

16. $-(-x)$ when $x = -23$

17. $-|20|$

18. $-(-x)$ when $x = |-3|$

Exercises 2.1

Fill in each blank with the appropriate symbol, either < or >.

1.   3___ -8

2. -3___ -8

3.   3 ___ 8

4. -9 ___ 5

5. -9 ___ -5

6.   9 ___ 5

7.   0___ -2

8.   3 ___ 0

9.   1 ___ -1

10. -10___ 10

With the following exercises, use the "$-$" symbol with one of the numbers to make the statement true.

11. $5 < 3$

12. $-5 > 8$

13. $-5 < -8$

14. $-5 > 3$

15. $1 > 1$

16. $-6 < -6$

17. $0 > 3$

18. $-6 > 0$

For the following exercises, fill in each blank with either "negative" or "the opposite of"

19. −3 is read as "_____ three."

20. −x is read as "_____ x."

21. −(−8) is read as "_____ _____ eight."

22. −(−w) is read as "_____ _____ w."

23. |−7| is read as "absolute value of _____ seven."

24. −8 is read as _____ eight."

25. − |x | is read as _____ absolute value of x."

26. −[−(−4)] is read as _____ _____ _____ four."

27. −w is read as "_____ _____ w."

28. − |−k | is read as" _____ absolute value of _____ k."

Evaluate the following expressions.

29. |−5|

30. −|5|

31. −(−4)

32. −|−7|

33. −x when x = 12

34. |−x | when x = 3

35. −|x | when x = 3

36. −x when x = −8

37. −|−x | when x = −8

38. −(−x) when x = −8

39. −x when x = 0

40. |0|

41. −|0|

42. −[−(−x)] when x = −1

Find the following number line distances.

43. distance between 2 and 7

44. distance between −2 and 7

45. distance between 2 and −7

46. distance between −2 and −7

47. distance between 0 and −13

48. distance between −13 and 0

49. distance between −3 and 3

50. distance between 19 and −19

7. |4| = 4 because the number line distance between 4 and 0 is 4.

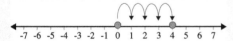

8. |−9| = 9 because the number line distance between −9 and 0 is 9.

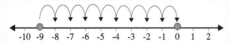

9. |1| = 1 because the number line distance between 1 and 0 is 1.

10. |−2| = 2 because the number line distance between −2 and 0 is 2.

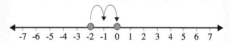

11. |−1| = 1 because the number line distance between −1 and 0 is 1.

12. |3| = 3 because the number line distance between 3 and 0 is 3.

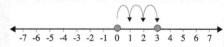

Practice C — Answers

13. $-(-5)$ reads "the opposite of the absolute value of 5." $-(-5) = 5$

14. $-|-5|$ reads "negative absolute value of negative 5." $-|-5| = -5$

15. $-x$ when $x = 23$ reads "the opposite of x when x is 23." $-x = -23$

16. $-(-x)$ when $x = -23$ reads "the opposite of the opposite of x when x is negative 23."
    $-(-x) = -(-(-23)) = -(23) = -23$

17. $-|20|$ reads "the opposite of the absolute value of 20." $-|20| = -20$

18. $-(-x)$ when $x = |-3|$ reads "the opposite of the opposite of x when x is the absolute value of negative 3"
    $-(-x) = -(-|-3|) = -(-3) = 3$

2.2 Addition and Subtraction

We've learned that addition represents the combined number of objects in two or more collections while subtraction represents the remaining number of objects in a collection after some have been removed.

With integers, however, there is increased complexity because numbers might be less than zero. It may be useful to think of an integer as a **net worth**.

There are two types of net worth:

Positive Net Worth

Positive net worth is when all of the money someone has is greater than all of the money they owe.

Negzative Net Worth

Negative net worth is when all of the money someone has is less than all of the money they owe.

A. Net Worth

Suppose that Chantelle's wallet contains the following items:

- An unpaid parking ticket for $35
- A birthday check from her Grandpa for $50
- An overdue book notice from the library showing an unpaid $5 fine
- A $20 bill

Some of these items represent money at her disposal that she can use to buy things. Other items represent money that she owes. If she has enough money at her disposal to pay off all of her debts, then she has a positive net worth of the amount that is left over. In this example, Chantelle has a total of $70 at her disposal, and she owes a total of $40. After all of her debts are paid off, she has $30 left, a positive net worth. Chantelle's net worth is represented by the integer 30.

Now suppose that Michael has these items in his wallet:

- $10 in cash
- An unpaid smartphone bill for $70
- A winning lottery ticket for $50

Michael has $60 at his disposal, and he owes $70. He does not have enough money to pay his debt completely off. If he pays as much as he can, he will still owe $10. So Michael has a negative net worth. His net worth is represented by the integer −10. In general, if a person does not have enough money at their disposal to pay off all of their debts, then their net worth is negative, it is the amount of debt left over when as much of it as possible has been paid off.

Suppose that Pedro has a net worth of $50. Then he walks out to his car and finds a $20 parking ticket under his windshield wiper for not parking with the nose of the car in the parking space. They always get you for that. He takes the parking ticket and puts it *into* his wallet. Because he's adding something into his wallet, we should think of this as an addition problem. His new net worth is $30 because he now has an additional $20 of debt to pay. Here is the integer addition that represents this scenario:

$$50 + (-20) = 30$$

Sue also has a net worth of $50. She decides to go watch a movie and have some snacks at the theater. To pay for all of this, she takes $20 *out* of her wallet. Because she's taking something out of her wallet, we can think of this as a subtraction problem. Her new net worth is $30 because she now has $20 less at her disposal. Here is the integer subtraction that represents this scenario:

$$50 - 20 = 30$$

Did you notice that in both of these scenarios the result was the same? Adding −20 and subtracting 20 have the same result.

Jim has a net worth of $50. He's walking down the street when he notices a $20 bill laying on the sidewalk. He picks it up and puts it *into* his wallet. His new net worth is $70:

$$50 + 20 = 70$$

Finally, Sylvia is another person with a net worth of $50. She currently has $70 in her wallet, but she also has a note in there from her Grandma reminding Sylvia that she still owes $20 because of her bet that vinegar wouldn't remove an ink stain from the sofa (it did). When Sylvia goes to visit her grandma, the old woman grabs Sylvia's wallet, takes the note *out* of the wallet, and says, "You don't have to pay up on that bet. I want you to have the money. I love you." Now Sylvia's net worth is $70 because she has $20 less debt to pay and a loving grandmother who knows a thing or two about vinegar. Here is the integer subtraction that represents this scenario:

$$50 - (-20) = 70$$

These four examples illustrate an important fact: adding an integer and subtracting the opposite of that integer have the same result. Knowing this, we can rewrite all subtraction problems as something called an **equivalent addition.** This means it's only necessary to develop rules for adding integers.

Equivalent Addition

Let a and b represent any two integers: $a + (-b) = a - b$ and $a + b = a - (-b)$

Example 1

Rewrite the following subtractions as equivalent additions.

1. $4 - 7$

2. $6 - (-2)$

Solutions

1. We don't change the first number. We change the operation from subtraction to addition and replace the second number with its opposite: $4 - 7$ becomes $4 + (-7)$.

2. $6 - (-2)$ becomes $6 + 2$.

Practice A

Re-write the subtractions as equivalent additions. Turn the page to check your answers.

1. $5 - 2$

2. $-8 - 3$

3. $4 - (-7)$

4. $-6 - (-9)$

5. $-2 - 3 - 4$

6. $3 - (-5) - 4$

B. Arrows on the Number Line

The number line gives us a way to visualize adding integers. We can think of integers as arrows on the number line. The direction of the arrow is the *sign* of the integer. Negative integers are arrows pointing to the left, and positive integers are arrows pointing to the right.

The length of the arrow is the *absolute value* of the integer. The integer -5, for example, is represented by an arrow that is 5 units long and points to the left. The integer 13 is represented by an arrow that is 13 units long and points to the right.

If a and b represent any two integers, we can use the number line to simplify $a + b$ by following these steps:

1. Place the a arrow so that it starts at 0 on the number line.

2. Place the b arrow so that it starts at the pointed end of the a arrow.

3. The pointed end of the b arrow is at the value of $a + b$ on the number line.

Example 2

Use the number line to simplify the following sums.

1. $6 + 4$

2. $-6 + (-4)$

3. $6 + (-4)$

4. $-6 + 4$

5. $3 + -7$

6. $-3 + 7$

Solutions

1. Both arrows point to the right. The second arrow ends at 10, so 6 + 4 = 10.

2. Both arrows point to the left. The second arrow ends at −10, so −6 + (−4) = −10.

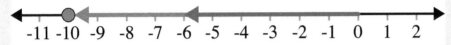

3. The arrows point in opposite directions. The second arrow ends at 2, so 6 + (−4) = 2.

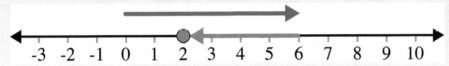

4. The arrows point in opposite directions. The second arrow ends at −2, so −6 + 4 = −2.

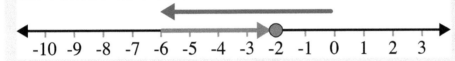

5. The arrows point in opposite direction. The second arrow ends at −4 so 3 + −7 = −4.

6. The arrows point in opposite direction. The second arrow ends at 4 so −3 + 7 = 4.

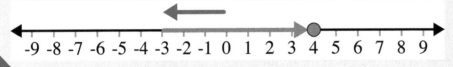

Practice A – Answers

1. 5 − 2 becomes 5 + (−2)
2. −8 − 3 becomes −8 + (−3)
3. 4 − (−7) becomes 4 + 7
4. −6 −(−9) becomes −6 + 9
5. −2 − 3 − 4 becomes −2 + (−3) + (−4)
6. 3 − (−5) − 4 becomes 3 + 5 + (−4)

Did you notice that in all of these examples, the direction of the longest arrow determines whether the sum is positive or negative? If the longest arrow points to the right, then the second arrow always ends to the right of zero on the number line. If the longest arrow points to the left, then the second arrow always ends to the left of zero on the number line.

The length of the arrow is the absolute value of the integer, so the longest arrow represents the term with the greatest absolute value. The direction of the longest arrow is the sign of that term. This gives us our first rule for adding integers.

Adding Integers — Rule 1

The sign of the sum $a + b$ is the same as the sign of whichever term, a or b, has the greatest absolute value.

For example, the sign of the sum $-2 + 5$ is positive because the absolute value of the positive number is greater than the absolute value of the negative number ($|5| > |-2|$). Similarly, the sign of $3 + (-9)$ is negative because the absolute value of the negative number is greater than the absolute value of the positive number ($|-9| > |3|$).

In the first two expressions from Example 2, the arrows point in the same direction, so we add their lengths to find the absolute value of $a + b$. In the last four expressions, the arrows point in the opposite direction, so we subtract their lengths to find the absolute value of $a + b$. This gives us the second rule for adding integers.

Adding Integers — Rule 2

If a and b have the *same sign*, the absolute value of $a + b$ is the *sum* of the absolute values of a and b.

If a and b have *opposite signs*, the absolute value of $a + b$ is the *difference* of the absolute values of a and b. In that case, we subtract the least absolute value from the greatest absolute value. The absolute value of $a + b$ cannot be negative.

Example 3

Determine if we need to add or subtract to find the absolute value of the following sums. Then find the sums.

1. $-4 + (-7)$
2. $-9 + 10$
3. $2 + 8$
4. $11 + (-3)$

Solutions

1. We would add because −4 and −7 have the *same* sign.

2. We would subtract because −9 and 10 have *opposite* signs.

3. We would add because 2 and 8 have the *same* sign.

4. We would subtract because 11 and −3 have *opposite* signs.

Practice B

Evaluate the following sums and differences. For differences, first rewrite the expression as the equivalent addition, and then evaluate as a sum. Turn the page to check your answers.

7. −8 + 2

8. −12 − (−5)

9. 5 − 8

10. −3 + (−7)

11. 5 − (−9)

12. 12 + (−13)

C. Properties of Addition

In Chapter 1, we learned that whole number addition has several important properties. All of these properties also apply to integer addition, and we gain a new property. Here is the full collection of properties for integer addition. If a, b and c represent any integer values, then:

> **commutative property**: $a + b = b + a$
>
> **associative property**: $(a + b) + c = a + (b + c)$
>
> **additive identity property**: $a + 0 = 0 + a = a$
>
> **additive inverse property**: $a + (−a) = − a + a = 0$

The last property doesn't make sense in the context of whole numbers, but it makes sense in the context of integers because we now have access to negative numbers. As we learned in Chapter 1, the first two properties allow us to add a string of numbers in whatever order we please. The result will be the same.

Example 4

Rewrite the following subtractions as equivalent additions. Then simplify the sum by using the properties of addition.

$7 + 5 − 3 + (−4) − (−8)$

Solution

$$7 + 5 - 3 + (-4) - (-8) = 7 + 5 + (-3) + (-4) + 8 \qquad \text{Rewrite subtractions as equivalent additions.}$$

$$= 7 + 5 + ((-3) + (-4)) + 8 \quad \text{Apply the associative property.}$$
$$= 7 + 5 + (-7) + 8$$

$$= 7 + 5 + (-7) + 8 \qquad\qquad \text{Apply the commutative property.}$$
$$= 7 + (-7) + 5 + 8$$

$$= 7 + (-7) + 5 + 8 \qquad\qquad \text{Apply the additive inverse property.}$$
$$= 0 + 5 + 8$$

$$= 0 + 5 + 8 \qquad\qquad\qquad \text{Apply the additive identity property,}$$
$$= 5 + 8 \qquad\qquad\qquad\qquad \text{and then add.}$$
$$= 13$$

It's important to note that none of these properties apply to the operation of subtraction. This is why it's so important to be able to rewrite subtractions as equivalent additions. As an operation, addition is fairly well behaved, but subtraction is not so well behaved. So we avoid it.

Practice C

Rewrite subtractions as equivalent additions. Then simplify the sum using the properties of addition. Turn the page to check your answer.

$-10 + 8 - 6 - (-2)$

Exercises 2.2

For the following exercises, rewrite the subtractions as equivalent additions.

1. $-5 - 9$
2. $-12 - 8$
3. $-23 - 22$
4. $-7 - 4$
5. $4 - (-7)$

6. $15 - (-2)$
7. $4 - (-35)$
8. $9 - (-4)$
9. $-11 - (-6)$
10. $-17 - (-5)$

11. $-6 - (-17)$
12. $-5 - (-13)$
13. $5 - 13 - (-3)$
14. $-8 - (-17) - 6$

For the following exercises, evaluate the expressions.

15. $8 + (-12)$
16. $-7 + (-13)$
17. $-17 + (-1)$
18. $18 + (-2)$
19. $-7 + 13$

20. $5 + (-11)$
21. $15 + (-1)$
22. $-17 + 1$
23. $-24 + (-16)$
24. $-24 + 16$

25. $-14 + 26$
26. $-14 + (-26)$
27. $-429 + 57$
28. $-5 + 11$

For the following exercises, rewrite any subtractions as equivalent additions. Then evaluate.

29. $-20 - 9$

30. $-7 - (-15)$

31. $-17 - (-5)$

32. $-2 - 90$

33. $8 - 12$

34. $-73 - 27$

35. $-7 - 37$

36. $1 - 82$

37. $82 - (-38)$

38. $375 - 823$

39. $872 - 884$

40. $22 - (-88)$

41. $-19 + (-8) - 2$

42. $7 - (-3) + (-2)$

43. $27 - (-53) + (-27) - 23$

44. $-49 + (-18) - 28 - (-49)$

For the following exercises, represent the net worth scenarios as integer sums. Then simplify the sums to calculate the net worth of each person.

45. Bob has the following items in his safe deposit box:

- an I-Owe-You note from James for $500

- a house promissory note for $40,000

46. Sally has the following items in her safe deposit box:

- a stock certificate worth $2000

- jewelry worth $500

- car loan paperwork with $2500 owed

47. Yau has the following items in his wallet:

- an unpaid electricity bill for $72

- a rare coin worth $54

- a $10 bill

- a note reminding him that he owes Sue $23

48. Sue has the following items in her wallet:

- an unpaid parking ticket for $27

- a note reminding her that Yau owes her $23

- a birthday check for $30

- an unpaid library fine for $6

Practice B — Answers

7. $-8 + 2 = -6$

8. $-12 - (-5) = -12 + 5 = -7$

9. $5 - 8 = 5 + (-8) = -3$

10. $-3 + (-7) = -10$

11. $5 - (-9) = 5 + 9 = 14$

12. $12 + (-13) = -1$

$$\xleftarrow{\hspace{2cm}}\ \underset{\substack{-12\,-11\,-10\ -9\ -8\ -7\ -6\ -5\ -4\ -3\ -2\ -1\ \ 0\ \ 1\ \ 2\ \ 3\ \ 4\ \ 5\ \ 6\ \ 7\ \ 8\ \ 9\ 10\ 11\ 12}}{\bullet}\ \xrightarrow{\hspace{2cm}}$$

49. Count gaps on the number line above to measure the distance between −9 and −2.

50. Count gaps on the number line above to measure the distance between 4 and −5.

51. Simplify the following: $|-9-(-2)|$

52. Simplify the following: $|4-(-5)|$

53. Let a and b represent any two integers. Studying the previous four problems, what do you think the value of $|a-b|$ represents?

54. Without counting gaps on a number line, what is the number line distance between −23 and 44?

55. Without counting gaps on a number line, what is the number line distance between −754 and 256?

56. Without counting gaps on a number line, what is the number line distance between −837 and 773?

Practice C — Answers

$$
\begin{aligned}
-10 + 8 - 6 - (-2) &= -10 + 8 + (-6) + 2 && \text{Rewrite subtraction as equivalent addition.}\\
&= -10 + 8 + 2 + (-6) && \text{Apply commutative property of addition.}\\
&= -10 + (8 + 2) + (-6) && \text{Apply associative property of addition.}\\
&= -10 + 10 + (-6)\\
&= 0 + (-6) && \text{Apply additive inverse property.}\\
&= -6
\end{aligned}
$$

2.3 Multiplication and Division

From the previous section, we know that we can represent integers as arrows on the number line. In this section, we will use this idea to help think about multiplication with integers, then we will develop rules for multiplying and dividing integers. We will also investigate exponent notation with integer-valued bases.

A. Using a Number Line to Understand Integer Multiplication

If a and b are any two integers, we can calculate the product of a and b ($a \cdot b$) by placing arrows representing the integer b end to end, starting at zero. If a is a positive integer, then we place the arrows representing b in the direction that the arrows point. The pointed end of the ath arrow is at the product of $a \cdot b$.

The picture illustrates how this works when b is 4 and a is 3. We place the arrows from left to right because the arrows representing 4 all point to the right. Notice that the product $3 \cdot 4 = 12$ is at the pointed end of the third arrow.

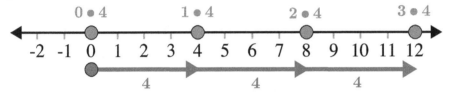

The next picture illustrates how this works when b is −4 and a is 3. In this case, we place the arrows from right to left because the arrows representing −4 all point to the left. We find the product $3 \cdot (-4) = -12$ at the pointed end of the third arrow.

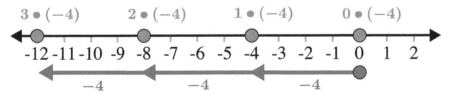

If a is a negative integer, then we place the arrows representing b in the opposite direction that the arrows point. The pointed end of the first arrow is at zero, and the non-pointed end of the last arrow is at the product $a \cdot b$. The number of arrows is the absolute value of a.

The picture below illustrates how this works when b is 4 and a is −3. We place 3 right-pointing arrows end to end, going to the left. The pointed end of the first arrow is at zero, and the non-pointed end of the third arrow is at −12, which is the product of −3 · 4. 3 · 4 is three arrows *after* zero, and −3 · 4 is three arrows *before* zero.

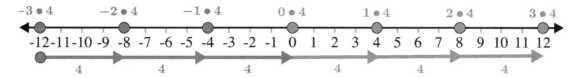

We can verify independently that $-3 \cdot 4 = -12$ because multiplication is commutative, which means that $-3 \cdot 4 = 4 \cdot (-3)$. If we place four arrows representing the integer -3 end to end, starting at zero and going to the left, the pointed end of the fourth arrow will be at -12, so $4 \cdot (-3) = -12$.

The picture below illustrates our final scenario, multiplying two negative numbers. When b is -4 and a is -3, we place three left-pointing arrows end to end, starting at zero and going to the *right* because the arrows representing b point to the *left* and a is a negative number. The non-pointed end of the third arrow is at 12, which is the product of $-3 \cdot (-4)$. Again, $3 \cdot (-4)$ is three arrows *after* zero, and $-3 \cdot (-4)$ is three arrows *before* zero.

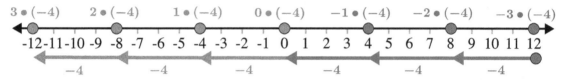

Practice A

Illustrate the following products by placing arrows on a number line. Then evaluate the products. Check your answers at the end of this section.

1. $5 \cdot 2$
2. $5 \cdot (-2)$
3. $-5 \cdot 2$
4. $-5 \cdot (-2)$
5. $-2 \cdot 3$
6. $-2 \cdot (-3)$

B. Rules for Multiplying Integers

We are ready to make some important observations about multiplying integers. Look at the second and third pictures above. If we multiply numbers that have *opposite* signs so that one of the numbers is positive and the other is negative, we place arrows to the left of zero. Either the arrows point to the left and we're placing them in the same direction, or the arrows point to the right and we're placing them in the opposite direction. This gives us our first rule for multiplying integers.

Multiplying Integers — Rule 1

Two numbers with different signs will always have a negative product.

Look at the first and fourth pictures above. In these scenarios, we multiply numbers with the *same* sign, either both are positive or both are negative. In both cases, we place the arrows to the right of zero. Either the arrows point to the right and we place them in the same direction, or the arrows point to the left and we place them in the opposite direction. This gives us a second rule.

Multiplying Integers — Rule 2

Two numbers with the same sign will always have a positive product.

When we use arrows to represent the product of $a \cdot b$, the number of arrows is the absolute value of a, and the length of each arrow is the absolute value of b. This means that the number line distance between zero and $a \cdot b$ is equal to the product of the absolute values of a and b.

Multiplying Integers — Rule 3

When a and b represent any integers, then $|a \cdot b| = |a| \cdot |b|$.

The properties of whole number multiplication that we learned in chapter 1 continue to apply to integer multiplication, but we have one new property, the **anti-identity property**. Let a, b, and c represent any integers, and the following are true:

- **commutative property:** $a \cdot b = b \cdot a$
- **associative property:** $(a \cdot b) \cdot c - a \cdot (b \cdot c)$
- **multiplicative identity property:** $a \cdot 1 = 1 \cdot a = a$
- **zero product property:** $a \cdot 0 = 0 \cdot a = 0$
- **anti-identity property:** $a \cdot (-1) = -1 \cdot a = -a$

Practice A — Answers

1. $5 \cdot 2 = 10$

2. $5 \cdot (-2) = -10$

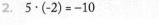

3. $-5 \cdot 2 = -10$

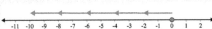

4. $-5 \cdot (-2) = 10$

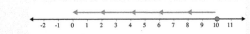

5. $-2 \cdot 3 = -6$

6. $-2 \cdot (-3) = 6$

Example 1

Simplify the following expressions.

1. $-2 \cdot 7$
2. $-8 \cdot (-3)$
3. $-5 \cdot (-2) \cdot (-3)$

Solutions

1. $-2 \cdot 7 = -14$ The integers have *different* signs, so the product is *negative*.

2. $-8 \cdot (-3) = 24$ The integers have the *same* signs, so the product is *positive*.

3. We can simplify this expression in two different ways, and both are correct:
    $(-5 \cdot (-2)) \cdot (-3) = 10 \cdot (-3) = -30$
    $-5 \cdot ((-2) \cdot (-3)) = -5 \cdot 6 = -30$
    We get the same answer either way. That shows you the associative property at work.

Practice B

Simplify the following expressions. Then check your answers at the end of this section.

7. $-6 \cdot 5$
9. $-2 \cdot (-8)$
11. $2 \cdot (-4) \cdot (-3)$

8. $6 \cdot (-5)$
10. $-4 \cdot 6 \cdot (-2)$
12. $-2 \cdot (-4) \cdot (-3)$

C. Rules for Dividing Integers

Next, let's adapt the definition of whole number division to apply to integers. If a and b are integers and b does not equal zero ($b \neq 0$), then the **quotient** of a and b is the integer c, provided that it exists, for which $a = b \cdot c$. In math symbols, that means $\frac{a}{b} = c$ if and only if $a = b \cdot c$. Consider the following four scenarios to find the rules for dividing integers:

1. a is *positive* and b is *positive*: If the quotient of a and b is defined, then there must be an integer c such that $a = b \cdot c$. In this case, c, the quotient of a and b, must be *positive* because it multiplies with the positive number b to give the positive result a.

2. a is *positive* and b is *negative*: If the quotient of a and b is defined, then there must be an integer c such that $a = b \cdot c$. In this case, c, the quotient of a and b, must be *negative* because it multiplies with the negative number b to give the positive result a.

3. a is *negative* and b is *positive*: If the quotient of a and b is defined, then there must be an integer c such that $a = b \cdot c$. In this case c, the quotient of a and b, must be *negative* because it multiplies with the positive number b to give the negative result a.

4. a is *negative* and b is *negative*: If the quotient of a and b is defined, then there must be an integer c such that $a = b \cdot c$. In this case, c, the quotient of a and b, must be *positive* because it multiplies with the negative number b to give the negative result a.

The rules for dividing integers are about the same as the rules for multiplying integers.

Dividing Integers

Let a and b be any two integers such that the quotient of a and b is defined.
If a and b have *opposite* signs, then their quotient is *negative*.
If a and b have the *same* sign, then their quotient is *positive*.

Example 2

Simplify the following quotients.

1. $\frac{-40}{-8}$ 2. $15 \div (-5)$ 3. $\frac{27}{0}$

Solutions

1. The dividend and divisor have the same sign—negative—so the quotient must be positive:
   $$\frac{-40}{-8} = 5$$

2. The dividend and divisor have opposite signs, so the quotient must be negative:
   $$15 \div (-5) = -3$$

3. The divisor is zero, so this quotient is undefined.

Practice C

Simplify the following quotients. Check your answers at the end of this section.

13. $\frac{-70}{10}$ 15. $\frac{0}{-13}$ 17. $\frac{-13}{0}$

14. $-24 \div (-6)$ 16. $\frac{70}{-10}$ 18. $\frac{-26}{-2}$

D. Exponent Notation with a Negative Base

We end this section by looking at repeated multiplication with a negative number. Consider the following examples: $(-2)^2, (-2)^3, (-2)^4,$ and $(-2)^5$. Below, we write them in expanded notation and then simplify them using rules for multiplying integers:

$$(-2)^2 = (-2) \cdot (-2) = 4$$
$$(-2)^3 = (-2) \cdot (-2) \cdot (-2) = 4 \cdot (-2) = -8$$
$$(-2)^4 = (-2) \cdot (-2) \cdot (-2) \cdot (-2) = -8 \cdot (-2) = 16$$
$$(-2)^5 = (-2) \cdot (-2) \cdot (-2) \cdot (-2) \cdot (-2) = 16 \cdot (-2) = -32$$

Notice that when the exponent is an *even number*, we end up with a *positive* answer, but when the exponent is an *odd number*, we end up with a *negative* answer. This is always true when the base is a negative number.

Exponents with a Negative Base

If a is a *negative* integer, and b is an *even* whole number, then a^b is *positive*.
If a is a *negative* integer and b is and *odd* whole number, then a^b is *negative*.

☠ Warning – Incorrect Approach ☠

-2^4 and $(-2)^4$ do not mean the same thing. The parentheses matter! $-2^4 = -16$, and $(-2)^4 = 16$. Here's the difference: -2^4 means "the opposite of 2^4." In other words, the negative symbol is not actually a part of the base. Remember the anti-identity property, that $-a = -1 \cdot a$. By applying this property, we get $-2^4 = -1 \bullet 2^4$. The correct order of operations to simplify this expression tells us that the exponent comes first:

$$-2^4 = -1 \cdot 2^4$$
$$= -1 \cdot 16$$
$$= -16$$

Practice D

Simplify the following quotients. Check your answers at the end of this section.

19. $(-5)^2$

20. -5^2

21. $(-5)^3$

22. $(-2)^6$

23. $(-2)^7$

24. -2^8

Exercises 2.3

For the following exercises, simplify the products and quotients.

1. $-4 \cdot 5$

2. $\frac{42}{-7}$

3. $\frac{-54}{-9}$

4. $-9 \cdot 6$

5. $3 \cdot (-2)$

6. $\frac{-18}{-3}$

7. $\frac{-39}{13}$

8. $8 \cdot (-8)$

9. $\frac{-49}{0}$

10. $-2 \cdot 5$

11. $-7 \cdot 7$

12. $\frac{-2}{0}$

13. $-8 \cdot (-3)$

14. $\frac{24}{0}$

15. $-2 \cdot (-12)$

16. $-13 \cdot (-3)$

17. $-5 \cdot 0$

18. $\frac{0}{-6}$

19. $0 \cdot 17$

20. $5 \cdot (-3) \cdot (-2)$

21. $3 \cdot (-2) \cdot (-7)$

22. $-5 \cdot (-8) \cdot (-3)$

23. $-2 \cdot (-3) \cdot (-4)$

24. $-4 \cdot (-1) \cdot 9 \cdot (-1)$

25. $-4 \cdot (-2) \cdot 10 \cdot (-3)$

26. $6 \cdot (-3) \cdot 5 \cdot (-2)$

27. $1 \cdot (-1) \cdot 1 \cdot (-1)$

28. $-2 \cdot (-3) \cdot 4 \cdot (-5)$

For the following exercises, simplify the expressions using the correct order of operations.

29. $(-3)^2$

30. $(-3)^3$

31. $(-2)^5$

32. $(-2)^4$

33. $(-3)^4$

34. -3^2

35. -2^4

36. $(-2)^6$

37. -3^4

38. $(5 - 8) \cdot (-6)$

39. $(-2 - 3) \cdot (-4)$

40. -2^6

41. $8 + 4 \cdot (-3)$

42. $9 - (-5) \cdot (-4)$

43. $-4 - (-1) \cdot (9)$

44. $7 - 3 \cdot (4)$

45. $\frac{-12 \cdot (-5 + (-11))}{-8}$

46. $\frac{-11 + 3 \cdot (-3)}{(2 - 7) \cdot (-2)}$

47. $\frac{4 - 8 \cdot (-9)}{(-6 + 25) \cdot (2)}$

48. $\frac{-22 \cdot (-7 - (-2))}{-10}$

49. $\frac{-9 + (-3)^2}{-3 \cdot (-5 + 2)}$

50. $\frac{-2 + 3(-3)}{-2(4) - (-2)^3}$

51. $\frac{-18 - 3 \cdot (-6)}{(4 - 12) \cdot (-4)}$

52. $\frac{3 - 4(-4)}{-4(4) + (-2)^4}$

53. $\frac{-2 + 3 + (-3)}{(4 - 4) \cdot (-2 - 7)}$

54. $\frac{12 + 3 \cdot (-4)}{(5 - 2) \cdot (-2)}$

2.4 Solving Equations

In Chapter 1, we used fact families to solve equations with whole numbers. To solve the equation $x + 7 = 19$, for example, we use the definition of subtraction which tells us that this equation is equivalent to $x = 19 - 7$. The two equations are in the same fact family. The complete solution to the equation looks like this:

$$x + 7 = 19$$
$$x = 19 - 7$$
$$x = 12$$

In this section, we'll develop a more sophisticated approach to solving equations that uses the **Addition Property for equations.**

A. Using the Addition Property to Solve Equations

One way to visualize an equation is to think of it as an old-fashioned scale with hanging baskets, as illustrated below. When the variable is replaced with a *solution* of the equation, then the scale becomes perfectly balanced. When the variable is replaced with a *non-solution* of the equation, the scale is not balanced.

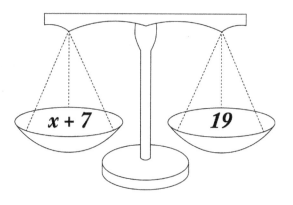

If we add exactly the same weight to both sides of a scale that's *balanced*, the scale stays balanced. If we add exactly the same weight to both sides of a scale that is *unbalanced*, the scale stays unbalanced. Applying this idea back to the equation, we get the **addition property for equations.**

Addition Property for Equations

When you add the *same number* to both sides of an equation, the resulting equation has exactly the same solution(s) as the original equation.

Let's see how we can use this principle to solve equations. We begin with the equation pictured in the drawing of the scale.

Example 1

Solve the following equations.

1. $x + 7 = 19$ **2.** $24 = w - 16$

Solutions

1.
$x + 7 = 19$	The original equation that we want to solve.
$x + 7 + (-7) = 19 + (-7)$	We add -7 to both sides to obtain and equation with exactly the same solution as the original equation.
$x + 0 = 12$	Then we simplify the integer sums.
$x = 12$	The number 12 is the only solution of this equation.

Then we check our solution.

$$12 + 7 = 19$$
$$19 = 19 \checkmark$$

2.
$24 = w - 16$	
$24 = w + (-16)$	Rewrite the subtraction as an equivalent addition.
$24 + 16 = w + (-16) + 16$	Add the opposite of -16 to both sides of the equation.
$40 = w + 0$	Simplify the integer sums
$40 = w$	The number 40 is the only solution to this equation.

Again, we check our solution.

$$24 = 40 - 16$$
$$24 = 24 \checkmark$$

The operation of subtraction has been rendered obsolete! We no longer need to use addition/subtraction fact families to solve equations. Instead, we can use the addition property. If the equation has any subtractions, we can just rewrite them as equivalent additions. Using integers instead of whole numbers makes our lives simpler because we have one less operation to worry about.

Practice A

Solve the following equations. Check your answers at the end of this section.

1. $x + 16 = 39$ 3. $z - 13 = -25$ 5. $447 = g - 214$

2. $18 = 26 + n$ 4. $-13 + h = 26$ 6. $301 = c + 71$

B. Solving Equations of the Type a − x = b

We can also use the addition property to solve equations in which the variable is the *minuend* of a difference. Consider the following scenario:

Example 2

After an intense afternoon of roller derby, Shawnelle was understandably thirsty and drank most of her 64-ounce Mountain Dew. Now she only has 27 ounces left. How many ounces did Shawnelle drink?

Solution

This situation can be modeled with

$$\left\{\begin{array}{c}\text{Initial amount}\\\text{of soda}\end{array}\right\} - \left\{\begin{array}{c}\text{Amount that}\\\text{Shawnelle drank}\end{array}\right\} = \left\{\begin{array}{c}\text{Amount of soda}\\\text{that remains}\end{array}\right\}$$

Let d represent the answer to the question — the number of ounces that Shawnelle drank. We know the initial amount of soda was 64 ounces, and now only 27 ounces remain. We plug 64, d, and 27 into our diagram and get the equation $64 - d = 27$.

We now solve this equation to answer the question.

$64 - d = 27$	This is the equation we want to solve.
$64 + (-d) = 27$	Rewrite the subtraction as an equivalent addition.
$64 + (-d) + d = 27 + d$	Add the opposite of $-d$ to both sides of the equation. We may not know the correct value for d, but we know that we're adding the same number to both sides of the equation, so the resulting equation has exactly the same solutions as the original equation.
$64 = 27 + d$	
$-27 + 64 = -27 + 27 + d$	Add −27 to both sides of the equation.
$37 = d$	Shawnelle drank 37 ounces of Mountain Dew. She was thirsty!

Now we verify our solution.

$$64 - 37 = 27$$
$$27 = 27 \checkmark$$

Practice B

Solve the following equations. Check your answers at the end of this section.

7. $9 - d = 10$

8. $-13 = 11 - m$

9. $19 - y = -5$

10. $-65 - e = -29$

11. $236 = -154 - p$

12. $167 - u = 221$

C. Solving Two-Step Equations

Now consider a slightly more complicated equation, $7x - 23 = 19$. This equation must be solved in two steps because the variable is a part of two operations. Suppose that we assign a number value to the variable x and then evaluate the expression $7x - 23$. According to the rules for the order of operations, which operation would we evaluate first? Neither of the operations, $(7 \cdot x)$ nor $(x - 23)$, are grouped, so we evaluate the multiplication first. Then we evaluate the subtraction.

To solve the equation, we want to end up with x alone on one side of the equal sign, so we must *undo* the operations. When we undo a sequence of operations, we undo them in the reverse order in which they were originally done.

To help visualize the process, consider the following analogy:

x	Pretend that the variable x is your bare foot.
$7x$	The first operation, multiplication, is putting on your sock, so $7x$ is your foot with your sock on it.
$7x - 23$	The second operation, subtraction, is you putting on your shoe. People laugh if you put on your socks after putting on your shoes. $7x - 23$ is your foot with your sock and shoe on.
$7x - 23 = 19$	To solve this equation, we must get x alone on one side of the equals sign, so we have to take off the sock and the shoe.
$\begin{aligned} 7x - 23 &= 19 \\ +23 \quad &+23 \\ \hline 7x &= 23 + 19 \\ 7x &= 42 \end{aligned}$	First, we take the shoe off. The second operation is subtraction, so we use the addition property to undo it. We end up with $7x$, which is your foot with your sock on.
$\begin{aligned} 7x &= 42 \\ \frac{42}{7} &= x \\ 6 &= x \end{aligned}$	Because the first operation is multiplication, we use the multiplication/division fact family to undo that operation. This is the second step, taking off the sock. We end up with x, which is your bare foot.
$\begin{aligned} 7(6) - 23 &= 19 \\ 42 - 23 &= 19 \\ 19 &= 19 \checkmark \end{aligned}$	And, of course, we check the solution because you wouldn't want to wear socks with holes in them. Make sure there aren't any holes in your socks and no mistakes in your solution!

In our next example, one of the operations is grouped. Since the grouped operation would be performed first, it must be undone last. Undoing the grouped operation is like taking off the sock in our previous analogy.

Example 3

Solve each equation.

1. $-45 = 9 \cdot (7 - x)$

2. $31 - 7k = -11$

Practice A — Answers

1.
$$x + 16 = 39$$
$$x + 16 + (-16) = 39 + (-16)$$
$$x = 23$$
Then we check our solution.
$$23 + 16 = 39$$
$$39 = 39 \checkmark$$

2.
$$18 = 26 + n$$
$$-26 + 18 = -26 + 26 + n$$
$$-8 = n$$
Then we check our solution.
$$18 = 26 + (-8)$$
$$18 = 18 \checkmark$$

3.
$$z - 13 = -25$$
$$z + (-13) = -25$$
$$z + (-13) + 13 = -25 + 13$$
$$z = -12$$
Then we check our solution.
$$-12 - 13 = -25$$
$$-12 + (-13) = -25$$
$$-25 = -25 \checkmark$$

4.
$$-13 + h = 26$$
$$13 + (-13) + h = 13 + 26$$
$$h = 39$$
Then we check our solution.
$$-13 + 39 = 26$$
$$26 = 26 \checkmark$$

5.
$$447 = g - 214$$
$$447 = g + (-214)$$
$$447 + 214 = g + (-214) + 214$$
$$661 = g$$
Then we check our solution.
$$447 = 661 - 214$$
$$447 = 447 \checkmark$$

6.
$$301 = c + 71$$
$$301 + (-71) = c + 71 + (-71)$$
$$230 = c$$
Then we check our solution.
$$301 = 230 + 71$$
$$301 = 301 \checkmark$$

Solution

1. $-45 = 9 \cdot (7 - x)$

For now, we still use multiplication/division fact families to undo any multiplications or divisions, but we can also use the addition property to undo any additions or subtractions.

First, we must undo the multiplication. If we substituted a number for x and simplified the expression $9 \cdot (7 - x)$, we would subtract first because of the parentheses and then multiply the result by 9. But, because we are undoing these operations, we must reverse the order, just like you take shoes and socks off in the *reverse order* of how you put them on.

$-45 = 9 \cdot (7 - x)$	The original equation.
$\dfrac{-45}{9} = 7 - x$	Replace with an equivalent equation from the same multiplication/division fact family.
$-5 = 7 - x$	Simplify the integer quotient.
$-5 = 7 + (-x)$	Rewrite the subtraction.
$-5 + x = 7 + (-x) + x$	Add x to both sides.
$-5 + x = 7$	Simplify.
$5 + (-5) + x = 5 + 7$	Add 5 to both sides.
$x = 12$	The number 12 is the only solution.

Now we verify our solution.

$$-45 = 9 \cdot (7 - 12)$$
$$-45 = 9 \cdot (-5)$$
$$-45 = -45 \checkmark$$

2. $31 - 7k = -11$

$31 + (-7k) = -11$	Rewrite the subtraction as an equivalent addition.
$-31 + 31 + (-7k) = -31 + (-11)$	Add -31 to both sides.
$-7k = -42$	Simplify the integer sums.
$k = \dfrac{-42}{-7}$	Use an equivalent equation from the same multiplication/division fact family.
$k = 6$	Simplify the integer quotient.

Now we verify our solution.

$$31 - 7 \cdot 6 = -11$$
$$31 - 42 = -11$$
$$-11 = -11 \checkmark$$

Practice B — Answers

7.

$$9 - d = 10$$
$$9 + (-d) = 10$$
$$9 + (-d) + d = 10 + d$$
$$9 = 10 + d$$
$$-10 + 9 = -10 + 10 + d$$
$$-1 = d$$

Then we check our solution.

$$9 - (-1) = 10$$
$$9 + 1 = 10$$
$$10 = 10 \checkmark$$

8.

$$-13 = 11 - m$$
$$-13 = 11 + (-m)$$
$$-13 + m = 11 + (-m) + m$$
$$-13 + m = 11$$
$$13 + (-13) + m = 13 + 11$$
$$m = 24$$

Then we check our solution.

$$-13 = 11 - 24$$
$$-13 = 11 + (-24)$$
$$-13 = -13 \checkmark$$

9.

$$19 - y = -5$$
$$19 + (-y) = -5$$
$$19 + (-y) + y = -5 + y$$
$$19 = -5 + y$$
$$5 + 19 = 5 + (-5) + y$$
$$24 = y$$

Then we check our solution.

$$19 - 24 = -5$$
$$19 + (-24) = -5$$
$$-5 = -5 \checkmark$$

10.

$$-65 - e = -29$$
$$-65 + (-e) = -29$$
$$-65 + (-e) + e = -29 + e$$
$$-65 = -29 + e$$
$$29 + (-65) = 29 + (-29) + e$$
$$-36 = e$$

Then we check our solution.

$$-65 - (-36) = -29$$
$$-65 + 36 = -29$$
$$-29 = -29 \checkmark$$

11.

$$236 = -154 - p$$
$$236 = -154 + (-p)$$
$$236 + p = -154 + (-p) + p$$
$$236 + p = -154$$
$$-236 + 236 + p = -236 + (-154)$$
$$p = -390$$

Then we check our solution.

$$236 = -154 - (-390)$$
$$236 = -154 + 390$$
$$236 = 236 \checkmark$$

12.

$$167 - u = 221$$
$$167 + (-u) = 221$$
$$167 + (-u) + u = 221 + u$$
$$-221 + 167 = -221 + 221 + u$$
$$-54 = u$$

Then we check our solution.

$$167 - (-54) = 221$$
$$167 + 54 = 221$$
$$221 = 221 \checkmark$$

Note that this example gives us an idea for an alternate process to solve the equation that we looked at in example 2. We will solve that equation again in example 4, thinking of it as a two-step equation.

Example 4

Solve $64 - d = 27$.

Solution

$$
\begin{aligned}
64 - d &= 27 & &\text{The equation we want to solve.} \\
64 + (-d) &= 27 & &\text{Rewrite the subtraction as an equivalent addition.} \\
-64 + 64 + (-d) &= -64 + 27 & &\text{Add } -64 \text{ to both sides.} \\
-d &= -37 & &\text{Simplify the integer sums.} \\
-1 \cdot d &= -37 & &\text{Use the anti-identity property to get the negative sign away from the variable.} \\
d &= \frac{-37}{-1} & &\text{Use an equivalent equation from the same multiplication/division fact family.} \\
d &= 37 & &\text{Simplify the integer quotient.}
\end{aligned}
$$

Now we verify our solution.

$$
\begin{aligned}
64 - 37 &= 27 \\
27 &= 27 \checkmark
\end{aligned}
$$

Practice C

Now it's your turn. Solve the following equations. Check your answers at the end of this section.

13. $7x - 25 = 24$

14. $-50 = 45 - 19d$

15. $-36 = (x - 12) \cdot 9$

16. $-13 - 5y = -33$

17. $55 = 19 - q$

18. $\frac{r-7}{3} = -22$

Exercises 2.4

Solve the following equations. Be sure to check your solutions!

1. $-5 + x = 12$

2. $r - 24 = -16$

3. $-27 = t - 6$

4. $-10 = r - 9$

5. $t - 45 = -31$

6. $n + (-14) = -20$

7. $25 = q + (-10)$

8. $42 = -67 - v$

9. $35 = s + 50$

10. $19 = w + 21$

11. $s + 73 = 48$

12. $16 + w = 70$

13. $v + (-22) = -15$

14. $39 - y = -21$

Which operation must be *undone* first. Use the Addition Property to undo the operation. Give the resulting equivalent equation.

15. $4c + 8 = 32$

16. $8x + 15 = -49$

17. $21b + (-17) = -17$

18. $-6x + (-13) = 65$

19. $8 + 9d = 35$

20. $3 + 5r = 48$

21. $39 = 93 + 54h$

22. $35 = 27 + 4t$

23. $-x + (-15) = -24$

24. $12 = -n + 13$

Which operation must be *undone* first? Choose the equation from the appropriate fact family that *undoes* this operation.

25. $7(r + 14) = 161$

26. $611 = 13(n - 23)$

27. $-99 = (23 + v) \cdot 3$

28. $42 = 2(h - 21)$

29. $14 = \dfrac{q - 23}{6}$

30. $\dfrac{r + 23}{5} = 6$

31. $2 = \dfrac{22}{4 + a}$

32. $\dfrac{63}{b + 3} = 21$

Solve the following equations. Be sure to check your solutions.

33. $4c + 8 = 32$

34. $8x + 15 = -49$

35. $21b - 17 = -17$

36. $-6x - 13 = 65$

37. $8 + 9d = 35$

38. $3 + 5r = 48$

39. $39 = 93 + 54h$

40. $35 = 27 + 4t$

41. $-x - 15 = -24$

42. $12 = -n + 13$

43. $7(r + 14) = 161$

44. $611 = 13(n - 23)$

45. $-99 = (23 + v) \cdot 3$

46. $42 = 2(h - 21)$

47. $14 = \dfrac{q - 23}{6}$

48. $\dfrac{r + 23}{5} = 6$

49. $2 = \dfrac{22}{4 + a}$

50. $\dfrac{63}{b + 3} = 21$

51. $3a - 7 = 26$

52. $31 = 2f - 15$

53. $94 = 7 - w$

54. $-23 - x = 404$

55. $-1 = \dfrac{n - 13}{5}$

56. $\dfrac{-72}{p + 4} = 8$

57. $(r - 7) \cdot 5 = 30$

58. $88 = 4(u + 13)$

59. $\dfrac{x}{3 - 4} = 5$

60. $41 = \dfrac{y}{8 + 9}$

Practice C — Answers

13.

$$7x - 25 = 24$$
$$7x + (-25) = 24$$
$$7x + (-25) + 25 = 24 + 25$$
$$7x = 49$$
$$x = \frac{49}{7}$$
$$x = 7$$

Now we verify our solution.
$$7 \cdot 7 - 25 = 24$$
$$49 - 25 = 24$$
$$24 = 24$$

14.

$$-50 = 45 - 19d$$
$$-50 = 45 + (-19d)$$
$$-45 + (-50) = -45 + 45 + (-19d)$$
$$-95 = -19d$$
$$\frac{-95}{-19} = d$$
$$5 = d$$

Now we verify our solution.
$$-50 = 45 - 19 \cdot 5$$
$$-50 = 45 - 95$$
$$-50 = 45 + (-95)$$
$$-50 = -50$$

15.

$$-36 = (x - 12) \cdot 9$$
$$\frac{-36}{9} = x - 12$$
$$-4 = x - 12$$
$$-4 = x + (-12)$$
$$-4 + 12 = x + (-12) + 12$$
$$8 = x$$

Now we verify our solution.
$$-36 = (8 - 12) \cdot 9$$
$$-36 = (8 + (-12)) \cdot 9$$
$$-36 = -4 \cdot 9$$
$$-36 = -36$$

16.

$$-13 - 5y = -33$$
$$-13 + (-5y) = -33$$
$$13 + (-13) + (-5y) = 13 + (-33)$$
$$-5y = -20$$
$$y = \frac{-20}{-5}$$
$$y = 4$$

Now we verify our solution.
$$-13 - 5 \cdot 4 = -33$$
$$-13 + (-5) \cdot 4 = -33$$
$$-13 + (-13) = -33$$
$$-33 = -33$$

17.

$$55 = 19 - q$$
$$55 = 19 + (-q)$$
$$-19 + 55 = -19 + 19 + (-q)$$
$$36 = -q$$
$$36 = -1 \cdot q$$
$$\frac{36}{-1} = q$$
$$-36 = q$$

Now we verify our solution.

$$55 = 19 - (-36)$$
$$55 = 19 + 36$$
$$55 = 55$$

18.

$$\frac{r-7}{3} = -22$$
$$r - 7 = -22 \cdot 3$$
$$r - 7 = -66$$
$$r + (-7) = -66$$
$$r + (-7) + 7 = -66 + 7$$
$$r = -59$$

Now we verify our solution.

$$\frac{-59 - 7}{3} = -22$$
$$\frac{-59 + (-7)}{3} = -22$$
$$\frac{-66}{3} = -22$$
$$-22 = -22$$

CHAPTER 3
Rational Numbers

What Are Numbers?

As we saw in Chapter 2, some counting scenarios require a new set of numbers. In that chapter, we expanded our study of whole numbers to include operations involving negative whole numbers. However, this set of numbers, the integers, is not sufficient to describe the solutions to many problems. As you probably realized, dividing one number by another does not always result in an integer.

For example, if we equally divide 3 dollars between 4 children, the amount of money each child receives won't be a whole dollar, but that doesn't mean they won't get some money from us. Similarly, an apple doesn't simply disappear when cut in half, and 2 people can easily share 1 apple.

When we ran into this problem in Chapter 1, you were instructed to respond "the quotient of a and b is not a whole number." This response does not deny the existence of the quotient. It merely acknowledges that our concept of numbers was not able to describe the quotient.

In this chapter, we will introduce a new and more comprehensive set of numbers, the set of rational numbers, which includes all quotients of integers with non-zero divisors.

3.1 Introduction to Rational Numbers

A. Rational Numbers

Let's look at an example that illustrates the need for the set of rational numbers. Consider the intrepid traveler, Ritu, who is preparing to hike the Oregon portion of the Pacific Crest Trail. She's preparing three food caches to leave at strategic points along the trail. She mixes 7 liters of Gatorade and divides it out evenly into 3 jugs, one for each cache. How many liters did Ritu put in each jug?

We can model the problem as follows. Let c represent the number of liters in each jug. Then the total number of liters of Gatorade is the product of 3 and c. We're told that the total number of liters of Gatorade is 7, so the multiplication equation $3c = 7$ models this situation. Here are the possible products of 3 and some whole number:

$$3 \cdot 0 = 0$$
$$3 \cdot 1 = 3$$
$$3 \cdot 2 = 6$$
$$3 \cdot 3 = 9$$
$$3 \cdot 4 = 12$$

Remember that we're looking for a way to multiple 3 by a whole number to get a product of 7. Note that if c is 2 or less, then $3c$ is *less* than 7, and if c is 3 or more, then $3c$ is *more* than 7. From this list, we can see that there is no whole number c such that $3c = 7$.

Does this mean that it's impossible for Ritu to divide 7 liters of Gatorade evenly into 3 jugs? Of course not. It just means that the number of liters in each jug is *more than* 2 liters and *less than* 3 liters. Because c represents the answer to a legitimate number question — "How many liters are in each jug?" — we should expect to find an exact number value for c. Simply answering "c is not a whole number" or the slightly better "c is more than 2 but less than 3" does not meet that standard. The rational number c is between two whole numbers.

In order to answer this question, we need to define the set of rational numbers.

Rational Numbers

Let a and b be any two integers. As long as $b \neq 0$ then $\frac{a}{b}$ is a rational number.

Rational numbers help us solve problems that do not have integer solutions. In fact, all of the integers are rational numbers, and we can write them as fractions above the number 1. The number 3 can be written $\frac{3}{1}$, -5 can be written as $\frac{-5}{1}$, and even 0 can be written as $\frac{0}{1}$.

The set of rational numbers contains the set of integers, just like the set of integers contains the set of whole numbers. Our definitions of numbers is becoming more and more sophisticated!

Example 1

Each of the following is a rational number. Some are also integers, but others are not. Identify which are integers and which are not integers. Each non-integer must be between two consecutive integers, so identify the consecutive integers on either side of the rational number.

1. $\dfrac{18}{3}$ 3. $\dfrac{17}{1}$ 5. $\dfrac{13}{-6}$

2. $\dfrac{-15}{5}$ 4. $\dfrac{17}{5}$

Solutions

1. $\dfrac{18}{3}$ is the integer 6.

2. $\dfrac{-15}{5}$ is the integer −3.

3. $\dfrac{17}{1}$ is the integer 17.

4. $\dfrac{17}{5}$ is not an integer. It is between 3 and 4 because 17 is between $5 \cdot 3$ and $5 \cdot 4$.

5. $\dfrac{13}{-6}$ is not an integer. It is between −3 and −2 because 13 is between −6(−3) and −6(−2).

Problem 3 in the previous example illustrates that every integer is also a rational number because the quotient of any integer a and 1 is always equal to a.

Problems 4 and 5 illustrate that even if rational numbers are not integers, they can still be either positive or negative. The rule for whether a rational number is positive or negative is the same as the rule for whether an integer quotient is positive or negative:

- If a and b have *opposite* signs, then the quotient of a and b is a *negative* rational number. The following are three equivalent ways to write the same negative rational number: $\dfrac{-3}{4} = \dfrac{3}{-4} = -\dfrac{3}{4}$. The negative symbol can be with the dividend, with the divisor, or in line with the fraction bar.

- If a and b have the *same* sign, then the quotient of a and b is a *positive* rational number. The following are two equivalent ways to write the same positive rational number: $\dfrac{8}{5} = \dfrac{-8}{-5}$.

Let's return to adventurous Ritu. We modeled the situation with the equation $3c = 7$, where c represents the number of liters of Gatorade in each jug. We can write an equivalent equation from the same multiplication/division fact family: $c = \dfrac{7}{3}$. The answer to our question is that c is the rational number $\dfrac{7}{3}$, so Ritu filled each bottle with $\dfrac{7}{3}$ liters of Gatorade!

Practice A

Read these story problems and use complete sentences describing the answer. Turn the page to check your solutions.

1. Seth must complete his math exam in 90 minutes. There are 23 questions on the exam. To avoid spending too much time on the early questions and thus running out of time to complete the exam, Seth decides to budget his time so that he has the same amount of time for each question. Which rational number represents the amount of time that Seth can spend on each question? This rational number is between which two consecutive integers?

2. Ben's Flowers is a business that takes care of flower boxes along the sidewalks in front of retail stores in town. Ben uses a cart with a water tank that holds 80 gallons to water the flowers each day during the peak of summer. If there are 33 flower boxes, which rational number represents the number of gallons Ben can dispense in each box to distribute the water evenly? This rational number is between which two consecutive integers?

3. Carolyn just bought a used car to commute to work. The previous owner claimed the car gets more than 25 miles per gallon. If Carolyn drives 230 miles in the first week and uses 9 gallons of gas, which rational number represents the number of miles driven for each gallon of gas used? This rational number is between which two consecutive integers? Does this support the previous owner's claim?

B. Fraction Notation

You may not have been very satisfied with the answer to Ritu's problem. After all, $\frac{7}{3}$ still looks more like a division problem than a number name. In fact, $\frac{7}{3}$ is *both* a quotient and a number name. This is an example of **fraction notation** for a rational number. Consider Figure 1:

To understand how to interpret $\frac{7}{3}$ as a number name, imagine dividing a single liter of Gatorade evenly into the three jugs. We would call the amount in each jug "a third of a liter." In Figure 1, the shaded portion of each jug illustrates how the first liter has been divided evenly between the three jugs. The amount in each jug is one third of a whole liter.

The subsequent liters are also evenly divided between the three jugs, so each jug contains 7 portions that are "a third of a liter." Another way to say this is "a total of 7 *thirds of a liter*." This is how we interpret the fraction notation $\frac{7}{3}$. We read that as "seven thirds."

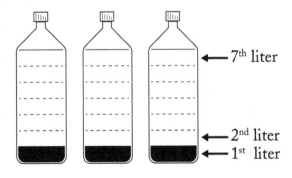

7th liter

2nd liter
1st liter

Figure 1

The numbers above and below the fraction line tell us how many portions the fraction represents and how much each portion is. The number *above* the fraction bar is called the **numerator**. It tells us how many portions the fraction represents. In our example, the numerator is 7, so the fraction represents 7 portions. The number *below* the fraction bar is called the **denominator**. It tells us how much each portion is. In our example, the denominator is 3, so each portion is one third of a whole.

Example 2

Read each fraction out loud. Then interpret the numerator and denominator of each fraction.

1. $\frac{3}{4}$

2. $\frac{314}{100}$

Solutions

1. The rational number $\frac{3}{4}$ is read as "3 fourths." It represents 3 portions, and each portion is one fourth of a whole.

2. The rational number $\frac{314}{100}$ is read as "314 hundredths." It represents 314 portions, and each portion is one hundredth of a whole.

Practice B

Write each rational number in fraction notation. Turn the page to check your answers.

4. Three fifths

5. Eighteen sevenths

6. Negative two elevenths

7. Twenty-three

8. Three two hundred forty fourths

9. Negative eight hundred fifty thirtheenths

C. Decimal Fractions

In Chapter 1, you learned the following rule:

> If n is a whole number, then 10^n can be evaluated by writing a one followed by n zeros.

Numbers like these are called "powers of ten." They are at the heart of our place value number system. In problem 2 of the last example, the fraction has a power of ten as its *denominator*. In that case, it can be written in decimal notation.

Decimal Fractions

Let a be an integer and n a whole number. Numbers in the form $\frac{a}{10^n}$ are called **decimal fractions**, and we can write them in decimal notation.

To write $\frac{a}{10^n}$ in decimal notation, we first write out the number a. Then we place a period, called a **decimal point**, between two numbers so that there are exactly n digits to the right of the decimal point. If the number a has fewer than n digits, we add as many zeros to the left of a as necessary so that there can be n digits to the right of the decimal point. If there are no non-zero digits to the *left* of the decimal point, then it is customary to write one zero to the left of the decimal point to draw attention to the decimal point (the number .01 should be written as 0.01).

Example 3

Convert the following numbers to decimal notation if they are written in fraction notation, or convert the numbers to fraction notation if they are written in decimal notation.

1. $\frac{314}{100}$ 2. $\frac{25}{1000}$ 3. 0.00059   4. 263.1

Solutions

1. $\frac{314}{100}$ is written as 3.14 in decimal notation. 100 is 10^2, so there should be two digits to the right of the decimal point.

2. $\frac{25}{1000}$ is written as 0.025 in decimal notation. 1000 is 10^3, so there should be three digits to the right of the decimal point. Note that we inserted one zero to the left of 25 in order to have three digits to the right of the decimal point. We added the zero to the *left* of the decimal point to draw attention to the decimal point.

3. 0.00059 is written as $\frac{59}{100,000}$ in fraction notation. There are five digits to the right of the decimal point, so the denominator must be $10^5 = 100,000$.

4. 263.1 is written as $\frac{2631}{10}$ in fraction notation. There is one digit to the right of the decimal point, so the denominator must be $10^1 = 10$.

Practice A — Answers

1. Seth should spend $\frac{90}{23}$ minutes on each question. The rational number $\frac{90}{23}$ is between the consecutive integers 3 and 4.

2. Ben can dispense $\frac{80}{33}$ gallons in each flower box to distribute the water evenly. The rational number $\frac{80}{33}$ is between the consecutive integers 2 and 3.

3. Carolyn drove the car $\frac{230}{9}$ miles for each gallon of gas used the first week. The rational number $\frac{230}{9}$ is between the consecutive integers 25 and 26. This supports the claim that the car gets more than 25 miles per gallon.

Practice C

Convert the following rational numbers into either fraction notation or into decimal notation. Turn the page to check your solutions.

10. 0.0039

12. 0.95

14. -0.1

11. $\frac{8}{10}$

13. $\frac{7319}{100}$

15. $-\frac{37,000}{10,000}$

Exercises 3.1

For the following exercises, determine whether the given rational number is an integer. If it is, state which integer. If it's not an integer, state the two consecutive integers that the rational number is between.

1. $\frac{27}{3}$

2. $\frac{35}{7}$

3. $\frac{42}{7}$

4. $\frac{63}{9}$

5. $\frac{37}{5}$

6. $\frac{48}{10}$

7. $\frac{58}{7}$

8. $\frac{64}{9}$

9. $\frac{-54}{9}$

10. $\frac{-51}{7}$

11. $\frac{66}{-11}$

12. $\frac{-52}{4}$

13. $\frac{0}{26}$

14. $\frac{32}{-8}$

15. $\frac{42}{-32}$

16. $\frac{28}{-14}$

17. $\frac{32}{-50}$

18. $\frac{0}{-14}$

For the following exercises, write each in fraction notation with integer numerators and denominators. Do not reduce.

19. The quotient of 34 and 17

20. The quotient of 15 and 74

21. The quotient of –2 and 3

22. The quotient of –11 and 24

23. –23 ÷ 26

24. 15 ÷ (–75)

25. 2 ÷ (–3)

26. –10 ÷ 20

27. 0.4

28. 0.33

29. –0.24

30. –2.1

31. 23.72

32. 5.791

33. –5.2

34. 0.3

35. 0.003

36. –0.00082

37. three eighths

38. five ninths

39. 13 sixteenths

40. 63 twentieths

41. four thirds

42. eight fifths

Practice B — Answers

4. $\frac{3}{5}$

5. $\frac{18}{7}$

6. $-\frac{2}{11}$

7. $\frac{23}{1}$

8. $\frac{3}{244}$

9. $-\frac{850}{13}$

For the following exercises, write each fraction in decimal notation.

43. $\frac{9}{10}$

44. $\frac{3}{10}$

45. $\frac{1}{10}$

46. $\frac{1}{100}$

47. $\frac{-19}{1000}$

48. $\frac{5473}{100}$

49. $-\frac{23}{10}$

50. $-\frac{4}{100}$

51. $-\frac{231}{10}$

52. $-\frac{2351}{1000}$

53. $\frac{1233}{10,000}$

54. $\frac{1}{1000}$

55. $\frac{1}{10,000}$

56. $-\frac{3}{10,000}$

57. $-\frac{835}{10}$

58. $\frac{993}{10,000}$

3.2 Divisibility

In this section, we develop the language and ideas needed to tell he difference between rational numbers that are integers and rational numbers that are not integers.

A. Introduction to Divisibility

In the previous section, you learned that a rational number in fraction notation $\frac{a}{b}$ may or may not be an integer. If $b \neq 0$ and $\frac{a}{b}$ is an integer, then we say that a is **divisible** by b. This means that the quotient of integer a and integer b results in an integer. When you check for **divisibility**, that's what you're trying to determine. The following phrases all mean the same thing:

- a is divisible by b
- a is a multiple of b
- b is a divisor of a
- b is a factor of a

They're all statements about divisibility. Let's look at some examples of how this works.

> **Example 1**
>
> 1. Is 15 divisible by 5?
> 2. Is 46 a multiple of 7?
> 3. Is 11 a divisor of 132?
> 4. Is 20 a factor of 4?
>
> **Solution**
>
> 1. *Yes,* 15 is divisible by 5. The fraction $\frac{15}{5}$ *is* an integer: $\frac{15}{5} = 3$
>
> 2. *No,* 46 is not a multiple of 7. The fraction $\frac{46}{7}$ is a rational number greater than 6 and less than 7, so it is not an integer.
>
> 3. *Yes,* 11 is a divisor of 132. The fraction $\frac{132}{11}$ *is* an integer: $\frac{132}{11} = 12$
>
> 4. *No,* 20 is not a factor of 4. The fraction $\frac{4}{20}$ is a rational number greater than 0 and less than 1, so it is not an integer. Note: 4 *is* a factor of 20, but 20 is *not* a factor of 4.

Here are a few important facts about divisibility. If n is any positive integer, then:

◆ n is divisible by 1. The fraction $\frac{n}{1}$ is equal to the integer n.

◆ n is divisible by n. The fraction $\frac{n}{n}$ is equal to the integer 1.

◆ n is *not* divisible by 0. The fraction $\frac{n}{0}$ is undefined. It is not equal to any integer.

◆ 0 is divisible by n. The fraction $\frac{0}{n}$ is equal to the integer 0.

◆ If n is *not* divisible by an integer a, n cannot be divisible by any multiple of a.

We work with fraction notation quite a bit for the rest of this book, and it will often be necessary for you to identify which numbers are — or aren't — divisors of a given number. The easiest test for divisibility is to use a calculator to check if integer a divided by integer b has an integer quotient, but you may not always have access to a calculator. Here are some ways to check to see whether an integer n is divisible by any integer between 1 and 10:

Is n divisible by 1?	Yes, always!
Is n divisible by 2?	If the ones digit of n is divisible by two, then so is n itself. In other words, if the ones digit is a 0, 2, 4, 6 or 8, then n is divisible by 2 no matter what the other digits are. The numbers 32, 114, and 4,316 are all divisible by 2 because the ones digits are all divisible by 2.
	Note: if n is divisible by 2 then we say that n is an **even number**. If n is *not* divisible by 2, we say that n is an **odd number**.
Is n divisible by 3?	Add up the digits of n. If the sum is divisible by 3, then so is n itself. If the sum is not divisible by 3 then neither is n itself. We can see that a small number like 192 is divisible by 3 because $1 + 9 + 2 = 12$ and $1 + 2 = 3$, and 3 is certainly divisible by itself, but this test works for large numbers, too! For example, is 7,592,838 divisible by 3? We calculate $7 + 5 + 9 + 2 + 8 + 3 + 8 = 42$. Next, $4 + 2 = 6$. Finally, 6 is divisible by 3 because 6 divided by 3 is 2, an integer. As a result, we know that 42 is divisible by 3, and that 7,592,838 is also divisible by 3.
Is n divisible by 4?	Add the sum of 2 times the tens digit *plus* the ones digit (2 · tens digit + ones digit). If the solution is divisible by 4, then so is n itself. If the answer is not divisible by 4, then neither is n itself. For example, is 7536 divisible by 4? Calculate $2 \cdot 3 + 6 = 12$. 12 is divisible by 4 because 12 divided by 4 is 3, an integer. Because 12 is divisible by 4, 7536 is also divisible by 4.
	Note: don't bother to do this test if you already know that n is not divisible by 2. Remember, if n is not divisible by 2, then n cannot be divisible by any multiple of 2.

Is *n* divisible by 5? If the ones digit of *n* is divisible by 5, then so is *n* itself. In other words, if the ones digit is either 0 or 5, then *n* is divisible by 5. Otherwise, *n* is not divisible by 5.

Is *n* divisible by 6? If *n* is divisible by both 2 and 3, then *n* is divisible by 6. Otherwise, *n* is not divisible by 6.

Is *n* divisible by 7? Subtract 2 times the ones digit from the remaining digits (All digits except ones − 2 · ones digit). If the answer is divisible by 7, then so is *n* itself. For example, is 3801 divisible by 7? Calculate $380 - 2 \cdot 1 = 378$. Is 378 divisible by 7? Calculate $37 - 2 \cdot 8 = 21$. We know that 21 is divisible by 7 because 21 divided by 7 is 3, an integer. Because 21 is divisible by 7, 378 is divisible by 7, we know 3801 is also divisible by 7.

Is *n* divisible by 8? Add 4 times the hundreds digit *plus* 2 times the tens digit *plus* the ones digit (4 · hundreds digit + 2 · tens digit + ones digit). You should get an answer between 0 and 63. If the answer is divisible by 8, then so is *n* itself. For example, is 23,584 divisible by 8? Calculate $4 \cdot 5 + 2 \cdot 8 + 4 = 40$. We know that 40 is divisible by 8 because 40 divided by 8 is 5, an integer. That means that 23,584 is also divisible by 8.

Note: don't bother doing this test if you already know that *n* is not divisible by 2 or 4. Remember, if *n* is not divisible by 2 or 4 then *n* cannot be divisible by any multiple of either 2 or 4.

Is *n* divisible by 9? Add up all the digits of *n*. If the answer is divisible by 9, then so is *n* itself. If the answer is not divisible by 9 then neither is *n* itself. This test can also be used to find out if the sum of digits is divisible by 9. For example, is 7,592,868 divisible by 9? Calculate $7 + 5 + 9 + 2 + 8 + 6 + 8 = 45$ and $4 + 5 = 9$, and 9 is divisible by 9, so 45 is divisible by 9 and 7,592,868 is also divisible by 9.

Note: don't bother to do this test if you already know that *n* is not divisible by 3. If *n* is not divisible by 3, then *n* cannot be divisible by any multiple of 3.

Is *n* divisible by 10? Look at the ones digit. If the one's digit is 0 then *n* is divisible by 10. Otherwise *n* is not divisible by 10.

Note: don't bother doing this test if you already know that *n* is not divisible by 2 or 5. Remember, if *n* is not divisible by 2 or 5 then *n* cannot be divisible by any multiple of either 2 or 5.

Example 2

Answer the following questions without using a calculator.

1. Which of the integers between 1 and 10 are divisors of 1995?
2. Which of the integers between 1 and 10 are divisors of 1656?

Solution

1. Which of the integers between 1 and 10 are divisors of 1995?

Because the one's digit, 5, is not divisible by 2, neither is 1995.

We won't bother doing tests for divisibility by 4, 6, 8 or 10 because 1995 cannot be divisible by any multiple of 2.

To test for divisibility by 3 and 9, we add up the digits: $1 + 9 + 9 + 5 = 24$. Because 24 is divisible by 3, so is 1995. Because 24 is not divisible by 9, neither is 1995.

To test for divisibility by 5 we look at the one's digit. Because the one's digit is 5, we know that 1995 is divisible by 5.

To test for divisibility by 7, we calculate $199 - 2 \cdot 5 = 189$ and $18 - 2 \cdot 9 = 0$. We know that 0 is divisible by 7, so that means is 189 and 1995 are, too.

Our answer then, is that the integers between 1 and 10 that are divisors of 1995 are 1, 3, 5 and 7.

2. Which of the integers between 1 and 10 are divisors of 1656?

The one's digit, 6, is even, so 1656 is also even — that is, divisible by 2.

To test for divisibility by 3 and 9, we add up the digits: $1 + 6 + 5 + 6 = 18$. Because 18 is divisible by both 3 and 9, so is 1656.

To test for divisibility by 4, we calculate $2 \cdot 5 + 6 = 16$. Because 16 is divisible by 4, so is 1656.

The one's digit is neither 0 nor 5, so 1656 is not divisible by 5. We won't bother to test for divisibility by 10 because 1656 is not divisible by 5.

1656 is divisible by both 2 and 3, so it is divisible by 6.

To test for divisibility by 7, we calculate $165 - 2 \cdot 6 = 153$ and $15 - 2 \cdot 3 = 9$. Because 9 is not divisible by 7, neither is 153. Because 153 is not divisible by 7, neither is 1656.

To test for divisibility by 8, we calculate $4 \cdot 6 + 2 \cdot 5 + 6 = 40$. Because 40 is divisible by 8, so is 1656.

Our answer, then, is that the integers between 1 and 10 that are divisors of 1656 are 1, 2, 3, 4, 6, 8 and 9.

Practice A

For each of the following numbers, list the integers between 1 and 10 that are divisors of the number. Then turn the page and check your answers.

1. 81
2. 41
3. 56
4. 240
5. 2548
6. 2809

B. Greatest Common Factor

In the last subsection, you learned that the two statements: "b is a factor of a" and "b is a divisor of a" both mean the same thing. We can use "factor" and "divisor" interchangeably. Here are all of the positive factors of each of the following numbers: 36, 24, 40.

- ◆ The factors (divisors) of 36 are 1, 2, 3, 4, 6, 9, 12, 18, 36
- ◆ The factors (divisors) of 24 are 1, 2, 3, 4, 6, 8, 12, 24
- ◆ The factors (divisors) of 40 are 1, 2, 4, 5, 8, 10, 20, 40

You may have noticed the pattern in each of these lists. The nth number from the left and the nth number from the right always multiply to the number for which you are finding divisors. For example, look at the divisors of 36:

$$1, 2, 3, 4, 6, 9, 12, 18, 36 \quad 1 \cdot 36 = 36$$
$$1, 2, 3, 4, 6, 9, 12, 18, 36 \quad 2 \cdot 18 = 36$$
$$1, 2, 3, 4, 6, 9, 12, 18, 36 \quad 3 \cdot 12 = 36$$
$$1, 2, 3, 4, 6, 9, 12, 18, 36 \quad 4 \cdot 9 = 36$$
$$1, 2, 3, 4, 6, 9, 12, 18, 36 \quad 6 \cdot 6 = 36$$

Now, compare the lists of divisors for 36 and 24. What is the greatest number in both lists?

Divisors of 36: 1, 2, 3, 4, 6, 9, 12, 18, 36

Divisors of 24: 1, 2, 3, 4, 6, 8, 12, 24

Notice that 12 is the greatest number in both lists. This is what we call the **greatest common factor** of 36 and 24. For short, we call the greatest common factor the "GCF." We write this in math symbols as GCF(36, 24) = 12.

Now look at the lists and evaluate GCF(24, 40). Did you come up with 8? If you didn't, try again.

We can find the GCF of a list of more than two numbers by looking for the greatest number that is a divisor of all the listed numbers. For example, GCF(36, 24, 40) = 4.

Example 3

Evaluate the following.
1. GCF(70, 126)
2. GCF(22, 66)
3. GCF(45, 52)

4. GCF(52, 70, 126)

Solution

1. Divisors of 70: 1, 2, 5, 7, 10, 14, 35, 70
Divisors of 126: 1, 2, 3, 6, 7, 9, 14, 18, 21, 42, 63, 126
GCF(70, 126) = 14

2. 22 is a divisor of both 22 and 66. Furthermore 22 can't possibly have any divisors greater than itself, so we conclude that GCF(22, 66) = 22.

3. Divisors of 45: 1, 3, 5, 9, 15, 45
Divisors of 52: 1, 2, 4, 13, 26, 52
GCF(45, 52) = 1

4. Divisors of 52: 1, 2, 4, 13, 26, 52
Divisors of 70: 1, 2, 5, 7, 10, 14, 35, 70
Divisors of 126: 1, 2, 3, 6, 7, 9, 14, 18, 21, 42, 63, 126
GCF(52, 70, 126) = 2

The problems in the previous example allow us to make a couple of important observations. Problem 2 illustrates a universal fact: If a is a divisor of b then GCF(a, b) = a. The third problem illustrates the least possible value for a GCF. Because 1 is a divisor of every integer, GCF(a, b) cannot be less than 1. If GCF(a, b) = 1, then we say a and b are **relatively prime**.

Practice B

Evaluate the following. Turn the page to check your answers.

7. GCF(9, 15)

8. GCF(72, 55)

9. GCF(18, 54)

10. GCF(45, 54)

11. GCF(8, 20, 52)

12. GCF(12, 21, 84)

Practice A — Answers

1. The integers between 1 and 10 that are divisors of 81 are 1, 3, and 9.

2. The integer between 1 and 10 that is a divisor of 41 is 1.

3. The integers between 1 and 10 that are divisors of 56 are 1, 2, 4, 7 and 8

4. The integers between 1 and 10 that are divisors of 240 are 1, 2, 3, 4, 5, 6, 8 and 10

5. The integers between 1 and 10 that are divisors of 2548 are 1, 2, 4, and 7.

6. The integer between 1 and 10 that is a divisor of 2809 is 1.

C. Least Common Multiple

Here are two different ways to ask the same question:

- ◆ Is 48 divisible by 12?

- ◆ Is 48 a multiple of 12?

In both cases, the answer is "yes," because $48 = 12 \cdot 4$. Similarly, 48 is a multiple of 8 because $48 = 8 \cdot 6$. Because 48 is a multiple of both 12 and 8, we say that it is a **common multiple** of 12 and 8. As with GCF, we use the word "common" to mean *shared* and not to mean *ordinary*. The number 12 has many multiples: 12, 24, 36, 48, 60, and 72 are just a few. The number 8 also has many multiples: 8, 16, 24, 32, 40, 48, 56, 64, and 72 are just a few. Any number on both of these lists is a common multiple of 12 and 8. Here are a few other common multiples of 12 and 8:

$24 = 12 \cdot 2$ and $24 = 8 \cdot 3$

$72 = 12 \cdot 6 =$ and $72 = 8 \cdot 9$

$0 = 12 \cdot 0$ and $0 = 8 \cdot 0$

$-48 = 12(-4)$ and $-48 = 8(-6)$

Note: there are common multiples of 12 and 8 which are less than 48. In fact, three appear in this partial list of common multiples: −48, 0, and 24.

If we include negative multiples in our discussion, then there is no smallest common multiple. If we ignore negative multiples then zero will always be the smallest common multiple of any two numbers. But if we only consider positive multiples, then 24 is the **least common multiple** of 12 and 8.

Least Common Multiple

If a and b are two non-zero integers, then the **least common multiple** of a and b, is the least *positive* integer that is a multiple of a and a multiple of b. We denote this with $\text{LCM}(a, b)$.

In the next section, we'll learn a technique to calculate $\text{LCM}(a, b)$ and $\text{GCF}(a, b)$ that is useful when a and b are large numbers, but for small numbers like 12 and 8, it's easier to list the multiples until you find a number on both lists.

Example 4

Evaluate $\text{LCM}(21, 12)$

Solution

The positive multiples of 21 are
    21, 42, 63, 84, 105, …
The positive multiples of 12 are
    12, 24, 36, 48, 70, 72, 84, 96, …
Since 84 is the smallest number on both lists, $\text{LCM}(21, 12) = 84$.

Practice C

Evaluate each least common multiple. Check your answers at the end of this section.

13. LCM(4, 6)   15. LCM(9, 12)   17. LCM(4, 3, 8)

14. LCM(5, 15)   16. LCM(24, 15)   18. LCM(3, 6, 42)

Exercises 3.2

Write out all of the divisors from 1 through 10 for each of the following numbers.

1.   18

2.   24   9.   1220   16.   2255

3.   11   10.   882   17.   2844

4.   13   11.   883   18.   1001

5.   15   12.   1000   19.   2345

6.   25   13.   10440   20.   2347

7.   48   14.   1855

8.   84   15.   2000

Find all of the positive divisors for each number.

21.   6   26.   63   31.   84

22.   8   27.   67   32.   45

23.   14   28.   71   33.   99

24.   16   29.   82   34.   55

25.   48   30.   75

Use the lists of divisors you made in exercises 21-32 to evaluate each GCF.

35.   GCF(6, 14)   38.   GCF(8, 63)   41.   GCF(6, 14, 48)

36.   GCF(8, 16)   39.   GCF(48, 67)   42.   GCF(75, 45)

37.   GCF(6, 48)   40.   GCF(63, 84)   43.   GCF (82, 84)

44. GCF(82, 63)   47. GCF(48, 84)   50. GCF(75, 63, 48)

45. GCF(14, 84)   48. GCF(63, 48)   51. GCF(14, 48, 84)

46. GCF(75, 82)   49. GCF(6, 48, 84)   52. GCF(45, 84, 63)

Evaluate each least common multiple.

53. LCM(6, 8)   57. LCM(18, 12)   61. LCM(12, 16, 15)

54. LCM(8, 12)   58. LCM(20, 15)   62. LCM(35, 28, 14)

55. LCM(7, 14)   59. LCM(6, 9, 12)   63. LCM(40, 30, 60)

56. LCM(9, 18)   60. LCM(3, 4, 9)   64. LCM(80, 60, 120)

Search online to find a divisibility test for the following numbers.

65. Find a divisibility test for 11. Write out a description of the divisibility test and illustrate how it works by testing if 1507 and 3679 are divisible by 11.

66. Find a divisibility test for 13. Write out a description of the divisibility test and illustrate how it works by testing if 1751 and 3679 are divisible by 13.

67. Find a divisibility test for 17. Write out a description of the divisibility test and illustrate how it works by testing if 1507 and 1751 are divisible by 17.

3.3 Prime Factorization

A. Primes and Composites

In Section 3.2, you learned some important divisibility facts. Here are two of them:

- If n is any positive integer, then n is divisible by n.
- If n is any positive integer, then n is divisible by 1.

When we combine these two facts, we can make the following observation:

> Let n be any integer greater than 1. Then n must have at least two *distinct* positive divisors (i.e. two positive divisors which are not equal to each other): 1 and n itself.

How do we know that 1 and n are distinct? We specified that n is an integer greater than 1, therefore n cannot be equal to 1. We will now use this observation to categorize all integers greater than 1. Those with exactly two positive divisors are in one category and those with more than two positive divisors are in the other category.

Composite Numbers and Prime Numbers.

If a positive integer has more than two positive divisors, then it is a **composite number**.
If a positive integer has exactly two positive divisors, then it is a **prime number**.
The number 1 is the only positive integer that is neither prime nor composite.

Our observation guarantees that this categorization is complete: every integer greater than 1 is either a prime or composite number. Only the number 1 itself does not fit into either category because it has only one positive divisor.

Here are all of the prime numbers between 2 and 100:

$$2, 3, 5, 7, 11, 13, 17, 19, 23, 29, 31, 37, 41, 43, 47, 53, 59, 61, 67, 71, 73, 79, 83, 89, 97$$

Any integer between 2 and 100 that is not on this list is a composite number. For example, 87 is an integer between 2 and 100 that is not in this list, so it must be composite. The positive divisors of 87 are 1, 3, 29, and 87. Because there are more than two, we confirm that 87 is composite.

What about numbers that are greater than 100? For example, if we want to determine whether 151 is prime, we could test every integer between 2 and 150 to find out if they are divisors of 151. If we find that none are, then we know that 151 is prime because 1 and 151 are the two distinct positive divisors. If any of those integers are divisors of 151, then we know that 151 is composite because there are at least three distinct positive divisors: 1, 151, and the divisor that we found between 2 and 150.

Before we jump in and do all of those 149 divisibility tests, which sure seems like a lot of work, maybe we should pause to consider whether we can get the correct result with fewer tests. It turns out that instead of testing every integer between 2 and 150, it's actually enough to test just these five prime integers: 2, 3, 5, 7, 11. We only need to do five trial divisions instead of 149! Here's why:

Trial Division

Let n be any integer greater than 1. If k is a positive number such that $k^2 > n$ and none of the prime numbers less than k are divisors of n then n must be a prime number.

Let's use 151 as our integer greater than one. Now we need a positive number k that is greater than 151 when squared. The smallest is 13, since 12 squared is 144, which is less than 151. 13 squared, however, is 169, and is therefore greater than 151. We found our two numbers ($13^2 = 169 > 151$), so we only have to test for divisibility by prime numbers that are less than 13. Those prime numbers are 2, 3, 5, 7 and 11. If any of these five is a divisor of 151, then 151 is composite. If none of these five are divisors of 151, then 151 is prime.

To see if the five prime numbers are divisors, enter the following divisions into your calculator to see if the quotient is an integer (you could do some of them with divisibility tests from Section 3.2 if you don't have your calculator):

$151 \div 2$

$151 \div 3$

$151 \div 5$

$151 \div 7$

$151 \div 11$

None of these quotients result in an integer, so 151 has no prime divisor less than 13. Because $13^2 > 151$, this proves that 151 is a prime number.

Example 1

Use trial division to determine whether each number is prime or composite.

1. 211   2. 221   3. 227

Solutions
In each case, it's enough to check for divisibility by the first six primes: 2, 3, 5, 7, 11, and 13 because the seventh prime, 17, squares to a number that is greater than any of these three: $17^2 = 289$. We find that:

1. 211 is *prime*.

2. 221 is *composite*: The positive divisors of 221 are 221, 17, 13, and 1.

3. 227 is *prime*.

Practice A

Use trial division to determine whether each number is prime or composite. Turn the page to check your answers.

1. 113

2. 117

3. 223

4. 228

5. 377

6. 379

B. Prime Factorization

A **factorization** of a number is an integer product that evaluates to that number. For example, the following are all factorizations of 24:

$$2 \cdot 12$$

$$3 \cdot 8$$

$$6 \cdot 4$$

$$2 \cdot 2 \cdot 2 \cdot 3$$

In all of these factorizations except the last one, at least one factor is a composite number. In the last factorization, all of the factors are prime. This product is called the *prime factorization* of 24. If you have a sharp eye, you may have noticed that we said *the* prime factorization instead of *a* prime factorization. That's because there is only one factorization of 24 in which all the factors are prime — as long as you don't count factorizations such as $2 \cdot 2 \cdot 2 \cdot 3$ and $2 \cdot 3 \cdot 2 \cdot 2$ as different. This is true for all integers greater than 1. This fact is important enough to have a name. It is called the **Fundamental Theorem of Arithmetic.**

Fundamental Theorem of Arithmetic

If n is any integer greater than 1, then there is one and only one factorization of n in which every factor is prime.

In a prime factorization, the factors may be written in any order, but it's customary to write them in order from least to greatest. That's why we write $2 \cdot 2 \cdot 2 \cdot 3$ instead of $2 \cdot 3 \cdot 2 \cdot 2$. We often use exponent notation to indicate repeated factors, too, so we write $2^3 \cdot 3$ instead of $2 \cdot 2 \cdot 2 \cdot 3$. We will call $2^3 \cdot 3$ the *condensed prime factorization* of 24.

Here are the steps you can follow to calculate the prime factorization of an integer n:

1. Divide out as many factors of 2 as possible.

2. When the remaining factor is not divisible by 2, divide out as many factors of 3 as possible.

3. When the remaining factor is not divisible by 3, divide out as many factors of 5 as possible.

4. Continue in this fashion, using prime divisors, until the remaining factor is itself prime.

If $p^2 > n$ and n is not divisible by any prime number less than p, then n is a prime number. In this case, the prime factorization of n is just n itself.

Example 2

Calculate the condensed prime factorization of 3300.

Solution

$3300 \div 2 = 1650$ Divide out as many twos as possible.

$1650 \div 2 = 825$ We can divide by two twice.

$825 \div 3 = 275$ Next, divide out as many threes as possible.

$275 \div 5 = 55$ Next, divide out as many fives as possible.

$55 \div 5 = 11$ We can divide by 5 twice.

11 We're left with 11, the largest prime factor.

This series of divisions reveals that:

$3300 = 2 \cdot 2 \cdot 3 \cdot 5 \cdot 5 \cdot 11 = 2^2 \cdot 3 \cdot 5^2 \cdot 11$

We use exponential notation for any factors that occur more than once. The factors 2 and 5 each occur twice, so they are showed being squared in the condensed prime factorization.

Practice B

Calculate the prime factorizations, presenting your answers as the condensed prime factorizations when factors repeat. Turn the page to check your solutions.

7. 52

8. 73

9. 750

10. 1000

11. 1155

12. 1900

C. Using Prime Factorizations to Evaluate GCF and LCM

When we introduced **greatest common factor** and **least common multiple**, you learned that you can evaluate the GCF of two numbers by listing out all of their positive divisors and looking for the greatest number on both lists. For LCM, you can start listing the multiples of both numbers until you find the first number that is on both lists of multiples. These techniques work well for finding the GCF or LCM of small numbers, but they become time-consuming when applied to larger numbers.

Fortunately, there is a more efficient technique for larger numbers that involves prime factorization. Here's how it works. We build the condensed prime factorization for GCF(a, b) by looking at the condensed prime factorizations of a and of b. Any prime factor of both a and b is also a prime factor of GCF(a, b). If your condensed prime factorization has prime numbers with exponents, then the exponent for the GCF prime factor is the *least* of the exponents for that prime factor in the condensed prime factorizations of a and of b.

Example 3

Evaluate GCF(7800, 8820)

Solution
First, calculate the condensed prime factorizations for 7800 and 8820. Include exponents of 1 as appropriate for clarity:

$7800 = 2^3 \cdot 3^1 \cdot 5^2 \cdot 13^1$
$8820 = 2^2 \cdot 3^2 \cdot 5^1 \cdot 7^2$

Since 2, 3, and 5 are prime factors of both 7800 and 8820 they must be prime factors of GCF(7800, 8820). Choose the least exponent for each of these prime factors. The condensed prime factorization of GCF(7800, 8820) is $2^2 \cdot 3^1 \cdot 5^1$, so:

GCF(7800, 8820) = $2^2 \cdot 3 \cdot 5$ = 60.

We build the condensed prime factorization of LCM(a, b) by including every prime factor of both numbers. For each prime factor, we chose the *greatest* exponent for that prime factor from the condensed prime factorizations of a and of b.

Example 4

Evaluate LCM(7800, 8820)

Solution
First we calculate the condensed prime factorizations for 7800 and 8820. We'll include exponents of 1 when that's appropriate for the sake of clarity.

$7800 = 2^3 \cdot 3^1 \cdot 5^2 \cdot 13^1$
$8820 = 2^2 \cdot 3^2 \cdot 5^1 \cdot 7^2$

We build the condensed prime factorization of LCM(7800, 8820) by choosing the greatest exponent for each prime factor, then evaluate the product of your list of prime factors to find the LCM:

LCM(7800, 8820) = $2^3 \cdot 3^2 \cdot 5^2 \cdot 7^2 \cdot 13^1$ = 1,146,600

Did you notice which factors of the condensed prime factorizations of a and b were left out in calculating LCM(7800, 8820)? They were the exact same factors that were used to calculate GCF(7800, 8820)! This is the case whenever you use this technique to calculate the GCF and LCM of two positive integers. This becomes a really interesting way to check your work:

If a and b are any two positive integers, then GCF(a, b) $\cdot$ LCM(a, b) = $a \cdot b$

You can check this on your calculator: 60 $\cdot$ 1,146,600 = 7800 $\cdot$ 8820!

Practice C

Use condensed prime factorizations to calculate the greatest common factors and least common multiples. Turn the page and check your solutions.

13. GCF(12, 54) and LCM(12, 54)

14. GCF(1400, 210) and LCM(1400, 210)

15. GCF(3300, 3800) and LCM(3300, 3822)

Exercises 3.3

For the following exercises, determine whether each number is prime or composite. If the number is composite, write one prime factor.

1.	13	3.	17	5.	8	7.	15
2.	9	4.	11	6.	12	8.	19

For the following exercises, use trial division to determine whether each number is prime or composite. If the number is composite, write one prime factor.

9.	127	12.	257	15.	491	18.	493
10.	131	13.	253	16.	299	19.	361
11.	301	14.	323	17.	433	20.	503

For the following exercises, calculate the condensed prime factorization for each number.

21.	22	25.	44	29.	135	33.	2744
22.	33	26.	72	30.	630	34.	2106
23.	63	27.	385	31.	6760	35.	4125
24.	50	28.	600	32.	304	36.	936

Practice A — Answers

1.	113 is *prime*	3.	223 is *prime*	5.	377 is *composite*
2.	117 is *composite*	4.	228 is *composite*	6.	379 is *prime*

Practice B — Answers

7.	$52 = 2^2 \cdot 13$	9.	$750 = 2 \cdot 3 \cdot 5^3$	11.	$1155 = 3 \cdot 5 \cdot 7 \cdot 11$
8.	73	10.	$1000 = 2^3 \cdot 5^3$	12.	$1400 = 2^3 \cdot 5^2 \cdot 7$

For the following exercises, use condensed prime factorizations to evaluate the GCF and LCM of each pair of numbers.

37. 63 and 135

38. 600 and 4125

39. 22 and 44

40. 33 and 44

41. 44 and 63

42. 63 and 50

43. 44 and 385

44. 6760 and 936

45. 63 and 385

46. 50 and 600

47. 385 and 4125

48. 385 and 936

49. 44 and 2744

50. 50 and 4125

51. 6760 and 4125

52. 72 and 600

53. 13 and 17

54. 11 and 19

55. 127 and 433
     Hint: Look at exercises 9 and 17.

56. 491 and 503
     Hint: Look at exercises 15 and 20.

Think about these exercises. And then do them, of course.

57. If two numbers are *relatively prime*, what is the easy way to evaluate their least common multiple? Hint: Look at exercises 41 and 42.

58. Which number is greater, GCF(a, b) or LCM(a, b)? Why?

59. If GCF(a, b) · LCM(a, b) = $a · b$, then what conclusion can you draw about a and b?

60. Does the condensed prime factorization technique for evaluating greatest common factors work for three or more numbers? Try using it to evaluate GCF(600, 6750, 936)

61. Does the condensed prime factorization technique for evaluating least common multiples work for three or more numbers? Try using it to evaluate LCM(600, 6760, 936).

62. At the end of Section 3.6, we learned that GCF(a, b) · LCM(a, b) = $a · b$. Is it also true that GCF(a, b, c) · LCM(a, b, c) = $a · b · c$? Use the previous two problems as a test.

63. Let a and b be positive integers and suppose that a is a multiple of b. What is LCM(a, b)?

64. Let a and b be positive integers and suppose that a is a multiple of b. What is GCF(a, b)?

Practice C — Answers

13. GCF(12, 54) = 6 and LCM(12, 54) = 108

14. GCF(40, 210) = 10 and LCM(40, 210) = 840

15. GCF(3300, 3822) = 6 and LCM(3300, 3822) = 2,102,100

3.4 Equivalent Fractions

In the set of integers, numbers have just one name. Differently named numbers are, in fact, different numbers. This is not the case in the set of rational numbers. In section 3.1, for example, we saw that $\frac{18}{3}$ is both a rational number and the integer, 6. That means that $\frac{18}{3}$ and 6 are different names for the same rational number.

   We do something similar with the names of people. An airplane pilot, Sid, may be known as "Captain Sid," "Sidney," "head honcho," or even "that guy piloting the plane," but he is still the same person regardless of how he is named. Rational numbers work the same way. $\frac{18}{3}$ and 6 are different number names for the same number.

A. Equivalent Fractions

After Manuel completed a grueling marathon, he was really hungry and ate a lot of pizza — the shaded portion of the pizza in Figure 1. How much pizza did he eat?

We can't use a whole number to answer the question about how much pizza Manuel ate. He did eat some pizza, so he ate more than zero pizzas, but there is some pizza left over, so he ate less than one pizza.

   This is an example of a number question that needs to be answered with a rational number. The solid black lines divide the pizza into 3 equal portions (thirds), and the shaded part covers two of those "one-third" portions. The number of pizzas Manuel ate is two thirds, or $\frac{2}{3}$. But we can see that the whole pizza was actually cut into 12 equal slices (twelfths),

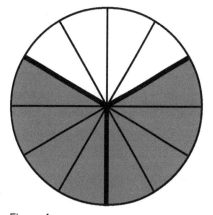

Figure 1

and Manuel ate 8 of them. So, another answer to the question is that the number of pizzas that Manuel ate is eight twelfths, or $\frac{8}{12}$. Both of these answers are correct! In fact, the two fractions given are different fraction notations for the same rational number. Any given rational number can be written as many different fractions. Two fractions that represent the same rational number are called **equivalent fractions.**

Equivalent Fractions

If $\frac{a}{b}$ is fraction notation for a rational number and c is any non-zero integer, then $\frac{a}{b}$ and $\frac{ac}{bc}$ are equivalent fractions. If two fractions are both equivalent to $\frac{a}{b}$, then they are equivalent to each other.

Figure 1 shows that each one-third portion consists of exactly four slices. That means that the part of the pizza that Manuel ate, 2 one-third portions, is two times four slices ($2 \cdot 4 = 8$ slices), while the whole pizza, 3 one-third portions, is three times four slices ($3 \cdot 4 = 12$ slices). With that information, we can make the following conversion between the two equivalent fractions:

$$\frac{2}{3} = \frac{2 \cdot 4}{3 \cdot 4} = \frac{8}{12}$$

Example 1

Explain why each pair of fractions is equivalent.

1. $\frac{40}{56}$ and $\frac{5}{7}$

2. $\frac{9}{12}$ and $\frac{15}{20}$

Solutions

1. $\frac{40}{56}$ and $\frac{5}{7}$ are equivalent fractions because when the numerator and denominator of $\frac{5}{7}$ are multiplied by the non-zero integer 8, the following conversion shows its equivalence with $\frac{40}{56}$.

    $$\frac{40}{56} = \frac{5 \cdot 8}{7 \cdot 8} = \frac{5}{7}$$

2. $\frac{9}{12}$ and $\frac{15}{20}$ are equivalent fractions because $\frac{9}{12}$ and $\frac{15}{20}$ are both equivalent to $\frac{3}{4}$, so they are equivalent to each other.

    $$\frac{3}{4} = \frac{3 \cdot 3}{4 \cdot 3} = \frac{9}{12} \qquad \frac{3}{4} = \frac{3 \cdot 5}{4 \cdot 5} = \frac{15}{20}$$

Practice A

Explain why each pair of fractions is equivalent. Turn the page to check your solutions.

1. $\frac{4}{10}$ and $\frac{2}{5}$

2. $\frac{6}{16}$ and $\frac{15}{40}$

3. $\frac{1}{3}$ and $\frac{9}{27}$

4. $\frac{15}{5}$ and $\frac{3}{1}$

5. $\frac{16}{32}$ and $\frac{23}{46}$

6. $\frac{65}{10}$ and $\frac{78}{12}$

B. Reducing to Lowest Terms

In Example 1, we see that in the fraction $\frac{5}{7}$, the numerator and denominator are relatively prime: $\text{GCF}(5, 7) = 1$. In the equivalent fraction $\frac{40}{56}$, the numerator and denominator are not relatively prime: $\text{GCF}(40, 56) = 8$. When the numerator and denominator of a fraction are relatively prime and the denominator is positive, we say that the fraction is in **lowest terms**. This means that there is no equivalent fraction that has a lower positive denominator.

When a fraction $\frac{a}{b}$ is in lowest terms, every equivalent fraction is of the form $\frac{ac}{bc}$ for some integer c. If $\frac{a}{b}$ is not in lowest terms, then there may be some equivalent fractions that do not have the form $\frac{ac}{bc}$ for any integer c. In the second problem from Example 1, there is no integer c such that $\frac{15}{20} = \frac{9c}{12c}$, because 15 is not divisible by 9. And yet, $\frac{9}{12}$ and $\frac{15}{20}$ are equivalent fractions because they are both equivalent to the fraction $\frac{3}{4}$.

When we **reduce** a fraction, we write an equivalent fraction in lowest terms. To reduce a fraction, we first evaluate the GCF of the numerator and denominator, then divide both the numerator and denominator by their GCF. Follow these steps to reduce a fraction:

If $\text{GCF}(a, b) = c$, and b is positive, then $\frac{a \div c}{b \div c}$ is in lowest terms.

If $\text{GCF}(a, b) = c$, and b is negative, then $\frac{a \div (-c)}{b \div (-c)}$ is in lowest terms.

Example 2

Reduce each fraction.

1. $\frac{45}{54}$

2. $\frac{24}{-32}$

Solutions

1. First, we find the GCF of the numerator and denominator: $\text{GCF}(45, 54) = 9$. The denominator is positive. We divide both the numerator and denominator by 9: $\frac{45}{54} = \frac{45 \div 9}{54 \div 9} = \frac{5}{6}$. The result is an equivalent fraction in lowest terms.

2. First, we find the GCF of the numerator and denominator: $\text{GCF}(23, -32) = 8$. This time, the denominator is negative, so we divide both the numerator and denominator by -8: $\frac{24}{-32} = \frac{24 \div (-8)}{-32 \div (-8)} = \frac{-3}{4}$. The result is an equivalent fraction in lowest terms.

Practice B

Reduce each fraction. Turn the page to check your solutions.

7. $\frac{7}{14}$

8. $\frac{72}{45}$

9. $\frac{-14}{21}$

10. $\frac{65}{55}$

11. $\frac{-49}{21}$

12. $\frac{75}{-30}$

C. Verifying Equivalent Fractions

How can you tell if two fractions are equivalent? One way is to reduce both fractions to their lowest terms. If you end up with the same answer, then the fractions are equivalent. For example, $\frac{-48}{80}$ and $\frac{21}{-35}$ both reduce to the same fraction: $-\frac{3}{5}$. It follows that they must be equivalent.

Another way is to perform cross multiplication on the pair of fractions. This results in two products that are equal if the fractions are equivalent, and different if they are not equivalent.

Verifying Equivalent Fractions

In general, if fractions are equivalent, then they must have the form $\frac{ac}{bc}$ and $\frac{ak}{bk}$ where a is any integer, b is a positive integer and c and k are both non-zero integers.

Note: multiplying the denominator of either fraction with the numerator of the other results in the same product, $abck$. These are called the **cross products** of the two fractions. If the cross products of two fractions are *equal*, then the fractions are equivalent. If the cross products are *not equal*, then the fractions are not equivalent.

$$\frac{ac}{bc} \diagdown\!\!\!\!\diagup \!\!= \frac{ak}{bk}$$

$$abck = abck$$

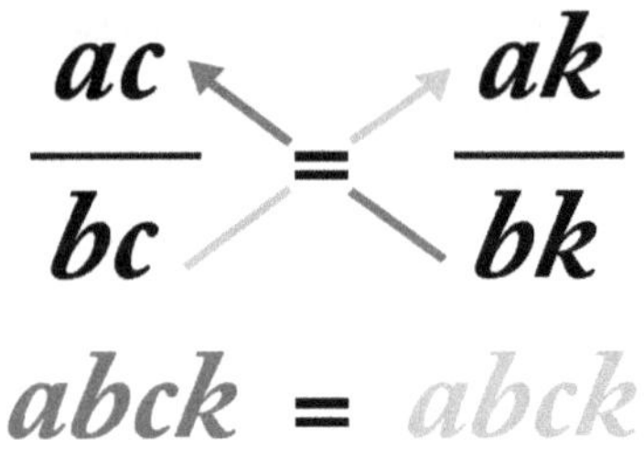

Figure 2

Example 3

Test each pair of fractions to find out if they are equivalent:

1. $\frac{-48}{80}$ and $\frac{21}{-35}$

2. $\frac{3}{10}$ and $\frac{10}{30}$

Solutions

1. The fractions are equivalent since the cross products are equal:

$$\frac{-48}{80} = \frac{21}{-35}$$

$$(-48)(-35) = 80 \cdot 21$$

$$1680 = 1680$$

2. The fractions are not equivalent since the cross products are not equal:

$$\frac{3}{10} \neq \frac{10}{30}$$

$$3 \cdot 30 \neq 10 \cdot 10$$

$$90 \neq 100$$

Practice C

Evaluate the cross products to find out if each pair of fractions are equivalent. Turn the page to check your solutions.

13. $\frac{3}{2}$ and $\frac{9}{6}$

14. $\frac{9}{6}$ and $\frac{15}{12}$

15. $\frac{15}{12}$ and $\frac{20}{16}$

16. $\frac{-16}{8}$ and $\frac{36}{-18}$

17. $\frac{6}{15}$ and $\frac{26}{65}$

18. $\frac{42}{30}$ and $\frac{60}{44}$

D. Designer Fractions

If we have one fraction, and we want to write a fraction with a specified denominator, we may or may not be able to do it. If the given fraction is in lowest terms, then it's a straightforward problem to check its equivalence with a specific denominator. Let $\frac{a}{b}$ be in lowest terms, and let c be an integer:

- If c is not a multiple of b, then an equivalent fraction with denominator c does not exist.

- If c is a multiple of b, then $c = b \cdot k$ for some integer, k. The fraction $\frac{a \cdot k}{c}$ is equivalent to $\frac{a}{b}$ because $\frac{a \cdot k}{c} = \frac{a \cdot k}{b \cdot k}$

Example 4

For the fraction $\frac{9}{5}$, do the following.

1. Find an equivalent fraction with a denominator of 35.

2. Find an equivalent fraction with a denominator of 17.

Solution

Since GCF(5, 9) = 1, the given fraction $\frac{9}{5}$ *is* in lowest terms. With that in mind, these are our solutions:

1. 35 is divisible by 5. We evaluate the number of times 35 is divisible by 5, $35 \div 5 = 7$, and calculate the value of the numerator when the denominator is 35: $\frac{9}{5} = \frac{9 \cdot 7}{5 \cdot 7} = \frac{63}{35}$.

The fractions are equivalent since the cross products of the two equivalent fractions are equal:

$$\frac{9}{5} = \frac{63}{35}$$
$$9 \cdot 35 = 5 \cdot 63$$
$$315 = 315$$

2. 17 is not divisible by 5, so no such equivalent fraction exists.

Practice A — Answers

1. $\frac{4}{10} = \frac{2 \cdot 2}{5 \cdot 2} = \frac{2}{5}$

2. $\frac{6}{16} = \frac{3 \cdot 2}{8 \cdot 2} = \frac{3}{8}$ and $\frac{15}{40} = \frac{3 \cdot 5}{8 \cdot 5} = \frac{3}{8}$

3. $\frac{1}{3} = \frac{1 \cdot 9}{3 \cdot 9} = \frac{9}{27}$

4. $\frac{15}{5} = \frac{3 \cdot 5}{1 \cdot 5} = \frac{3}{1}$

5. $\frac{16}{32} = \frac{1 \cdot 16}{2 \cdot 16} = \frac{1}{2}$ and $\frac{23}{46} = \frac{1 \cdot 23}{2 \cdot 23} = \frac{1}{2}$

6. $\frac{65}{10} = \frac{13 \cdot 5}{2 \cdot 5} = \frac{13}{2}$ and $\frac{78}{12} = \frac{13 \cdot 6}{2 \cdot 6} = \frac{13}{2}$

Practice B — Answers

7. $\frac{1}{3}$

8. $\frac{8}{5}$

9. $\frac{-2}{3}$

10. $\frac{13}{11}$

11. $\frac{-7}{3}$

12. $\frac{5}{2}$

In the previous example, we started with a fraction already in lowest terms, but problems won't always be that polite. If we are given a fraction that is not in lowest terms, then we must reduce the given fraction to lowest terms before finding its equivalent.

Example 5

Find a fraction equivalent to $\frac{30}{18}$ with denominator 24.

Solution

Even though 24 is not divisible by 18, this may still be possible because $\frac{30}{18}$ is not in lowest terms. To reduce $\frac{30}{18}$, first evaluate GCF(30, 18) = 6. Then, calculate $\frac{30 \div 6}{18 \div 6} = \frac{5}{3}$. Notice that 24 is divisible by 3. In fact, 24 ÷ 3 = 8. So, the fraction $\frac{5 \cdot 8}{3 \cdot 8} = \frac{40}{24}$ is equivalent to $\frac{30}{18}$.

Practice D

Find equivalent fractions with the specified denominators. If the given fraction is not in lowest terms, then reduce it first. Turn the page and check your solutions.

19. Find the fraction equivalent to $\frac{3}{4}$ that has a denominator of 52.

20. Find the fraction equivalent to $\frac{3}{4}$ that has a denominator of 26.

21. Find the fraction equivalent to $\frac{27}{63}$ that has a denominator of 14.

22. Find the fraction equivalent to $\frac{10}{22}$ that has a denominator of 55.

23. Find the fraction equivalent to $\frac{6}{8}$ that has a denominator of 14.

24. Find the fraction equivalent to $\frac{72}{45}$ that has a denominator of 15.

Exercises 3.4

If the fraction is not in lowest terms in the following exercises, reduce it. If the fraction *is* in lowest terms, write "already reduced."

1.	$\frac{4}{6}$	4.	$\frac{3}{12}$	7.	$\frac{76}{57}$	10.	$\frac{34}{12}$
2.	$\frac{6}{8}$	5.	$\frac{18}{12}$	8.	$\frac{44}{-14}$	11.	$\frac{27}{34}$
3.	$\frac{8}{12}$	6.	$\frac{-35}{42}$	9.	$\frac{39}{-52}$	12.	$\frac{54}{68}$

13. $\frac{25}{-16}$ 16. $\frac{-65}{52}$ 19. $\frac{21}{12}$ 22. $\frac{24}{42}$

14. $\frac{-5}{8}$ 17. $\frac{81}{18}$ 20. $\frac{12}{31}$

15. $\frac{-7}{28}$ 18. $\frac{9}{-14}$ 21. $\frac{-41}{7}$

Evaluate the cross products for each pair of fractions to find out if they are equivalent.

23. $\frac{4}{7}$ and $\frac{12}{21}$ 28. $\frac{12}{17}$ and $\frac{18}{23}$ 33. $\frac{-14}{20}$ and $\frac{28}{-40}$

24. $\frac{5}{12}$ and $\frac{25}{60}$ 29. $\frac{-3}{12}$ and $\frac{6}{-24}$ 34. $\frac{12}{15}$ and $\frac{24}{30}$

25. $\frac{4}{14}$ and $\frac{24}{70}$ 30. $\frac{-23}{31}$ and $\frac{92}{-124}$ 35. $\frac{55}{80}$ and $\frac{143}{208}$

26. $\frac{8}{12}$ and $\frac{10}{30}$ 31. $\frac{72}{-27}$ and $\frac{-105}{40}$ 36. $\frac{115}{65}$ and $\frac{209}{119}$

27. $\frac{-15}{8}$ and $\frac{105}{-56}$ 32. $\frac{-42}{91}$ and $\frac{18}{-39}$

For the following exercises, write a fraction equivalent to the given fraction with the specified denominator. If no such fraction exists, say so.

37. Fraction: $\frac{4}{9}$ Denominator: 27   40. Fraction: $\frac{3}{7}$ Denominator: 14

38. Fraction: $\frac{-3}{7}$ Denominator: −28   41. Fraction: $\frac{8}{13}$ Denominator: −43

39. Fraction: $\frac{-2}{5}$ Denominator: 10   42. Fraction: $\frac{24}{40}$ Denominator: 24

Practice C — Answers

13. The cross products are $3 \cdot 6 = 18$ and $2 \cdot 9 = 18$. The cross products are equal, so the fractions are equivalent.

14. The cross products are $9 \cdot 12 = 108$ and $6 \cdot 15 = 90$. The cross products are *not* equal, so the fractions are *not* equivalent.

15. The cross products are $15 \cdot 16 = 240$ and $12 \cdot 20 = 240$. The cross products are equal, so the fractions are equivalent.

16. The cross products are $-16 \cdot (-18) = 288$ and $8 \cdot 36 = 288$. The cross products are equal, so the fractions are equivalent.

17. The cross products are $6 \cdot 65 = 390$ and $15 \cdot 26 = 390$. The cross products are equal, so the fractions are equivalent.

18. The cross products are $42 \cdot 44 = 1848$ and $30 \cdot 60 = 1800$. The cross products are *not* equal, so the fractions are *not* equivalent.

43. Fraction: $\frac{4}{13}$ Denominator: 39

44. Fraction: $\frac{9}{4}$ Denominator: 24

45. Fraction: $\frac{-36}{27}$ Denominator: –12

46. Fraction: $\frac{-23}{17}$ Denominator: –26

47. Fraction: $\frac{45}{20}$ Denominator: 16

48. Fraction: $\frac{27}{18}$ Denominator: 21

49. Fraction: $\frac{10}{4}$ Denominator: 15

50. Fraction: $\frac{-14}{27}$ Denominator: –44

51. Fraction: $\frac{42}{-70}$ Denominator: 45

52. Fraction: $\frac{96}{-36}$ Denominator: –21

3.5 Multiplication and Division

Multiplication was introduced in Chapter 1 as a shorthand notation for repeated addition. This interpretation works well for whole number multiplication and is easily adapted to describe integer multiplication, too.

However, it takes a pretty big mental stretch to use this interpretation of multiplication to describe multiplying rational numbers. After all, the product of $\frac{17}{3}$ and $\frac{5}{11}$ cannot be easily visualized. Instead, we need a different interpretation of multiplication.

A. Area of a Rectangle

In Chapter 1, you learned that a rectangle measuring l units long and w units wide has an area of $l \cdot w$ square units. This is true when l and w represent whole numbers, but it's also true when l and w represent non-negative rational numbers.

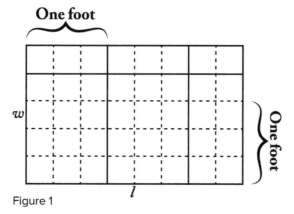

Figure 1

In Figure 1, solid *vertical* lines divide the rectangle's length into feet. The vertical dotted lines divide each foot of the rectangle's length into equal portions of one third of a foot. Because we can count eight of these one-third foot portions that span the length of the rectangle, the length of the rectangle is $l = \frac{8}{3}$ feet.

The horizontal solid lines also divide the rectangle's width into feet. The horizontal dotted lines divide each foot of the rectangle's width into four equal portions of one fourth of a foot. Because we can count five of these one-fourth foot portions that span the width of the rectangle, the width of the rectangle is $w = \frac{5}{4}$ feet.

Now we can use multiplication to calculate the area of the rectangle: $l \cdot w = \frac{8}{3} \cdot \frac{5}{4}$ square feet. Notice that the dotted lines divide each one foot by one foot square into 12 equal rectangles — three columns of four rectangles each. That means that the area of each of these little rectangles is one twelfth of a square foot. The entire big rectangle is covered perfectly by 40 of these little rectangles, eight columns of 5 rectangles each. That means the area of the big rectangle is $\frac{40}{12}$ square feet. We now have two calculations for the area of this rectangle. Because both are correct, they must be equal to each other: $l \cdot w = \frac{8}{3} \cdot \frac{5}{4} = \frac{40}{12}$.

Let's take a closer look at how this multiplication works. The *numerator* of the product, 40, represents the total number of little rectangles packed into the big rectangle. To count these little rectangles, we multiplied the numerators of the length and width: $8 \cdot 5 = 40$. The *denominator* of the product, 12, represents the number of little rectangles in each one foot by one foot square. To count the little rectangles in one square foot, we multiplied the denominators of the length and width: $3 \cdot 4 = 12$. This example illustrates the following rule for multiplying fractions.

Multiplying Rational Numbers in Fraction Notation

If $\frac{a}{b}$ and $\frac{c}{d}$ are rational numbers, then $\frac{a}{b} \cdot \frac{c}{d} = \frac{ac}{bd}$.

The *numerator* of the product is the product of the *numerators* of each factor. The *denominator* of the product is the product of the *denominators* of each factor.

The fractions that we used for the length and width of the rectangles were both in lowest terms:

$$GCF\,(8, 3) = 1$$

$$GCF\,(5, 4) = 1$$

The fraction that represents the product, however, is not in lowest terms:

$$GCF\,(40, 12) = 4$$

This common factor of 4 comes from the numerator of one factor, $\frac{8}{3}$, and the denominator of the other factor, $\frac{5}{4}$. After we evaluate a product we may wish to reduce it to lowest terms:

$$\frac{40}{12} = \frac{4 \cdot 10}{4 \cdot 3} = \frac{10}{3}$$

Example 1

Evaluate each product. If the product is not in lowest terms, reduce it.

1. $\frac{3}{5} \cdot \frac{10}{7}$
2. $\frac{-4}{9} \cdot \frac{-2}{3}$
3. $\frac{2}{5} \cdot \frac{5}{2}$

Solutions

1.
$$\frac{3}{5} \cdot \frac{10}{7} = \frac{30}{35}$$
$$= \frac{6 \cdot 5}{7 \cdot 5}$$
$$= \frac{6}{7}$$
Multiply the numerators and denominators, then reduce to lowest terms.

2.
$$\frac{-4}{9} \cdot \frac{-2}{3} = \frac{8}{27}$$
Multiply the numerators and denominators. The resulting fraction is already in lowest terms.

3.
$$\frac{2}{5} \cdot \frac{5}{2} = \frac{10}{10}$$
$$= 1$$
Multiply the numerators and denominators, then reduce to lowest terms.

Practice A

Evaluate each product. If the product is not in lowest terms, reduce it. Turn the page to check your solutions.

1. $\frac{2}{3} \cdot \frac{5}{7}$

2. $\frac{1}{3} \cdot \frac{6}{5}$

3. $\frac{15}{-8} \cdot \frac{-4}{5}$

4. $\frac{10}{7} \cdot \frac{13}{5}$

5. $-\frac{6}{13} \cdot \frac{5}{12}$

6. $\frac{23}{6} \cdot \frac{6}{23}$

B. Reciprocals

You may have noticed that in the previous example and practice, the last problems reduced to the integer 1. In both cases, the factors were **reciprocals** of each other.

If two rational numbers are reciprocals of each other, their product is always equal to 1. In the same way, if the product of two rational numbers is equal to 1, then they must be reciprocals of each other. Reciprocals are also called "**multiplicative inverses.**"

Reciprocals

Let p and q be rational numbers.
If $p \cdot q = 1$, then p and q are reciprocals.
If p and q are reciprocals, then $p \cdot q = 1$.

Every rational number except for zero has a unique reciprocal. It is important to remember, however, that fraction notations for rational numbers are not unique. Therefore it is possible for two fractions to be reciprocals even if they don't have the forms $\frac{a}{b}$ and $\frac{b}{a}$. We can always test whether fractions are reciprocals by evaluating their product. If the product is 1, then the fractions are reciprocals. If the product is not 1, then the fractions are not reciprocals.

Example 2

Test whether the following pairs of fractions are reciprocals.

1. $\frac{4}{8}$ and $\frac{4}{2}$
2. $\frac{-3}{5}$ and $\frac{5}{3}$
3. $\frac{-2}{7}$ and $\frac{-7}{2}$

Solutions

1. $\frac{4}{8} \cdot \frac{4}{2} = \frac{16}{16} = 1$, so these fractions are reciprocals.
   Note that they can both be reduced: $\frac{4}{8} = \frac{1}{2}$ and $\frac{4}{2} = 2$.

2. $\frac{-3}{5} \cdot \frac{5}{3} = \frac{-15}{15} = -1$, so these fractions are not reciprocals.

Note that reciprocals must have the same sign, either both positive or both negative, since their product, 1, is positive.

3. $\frac{-2}{7} \cdot \frac{-7}{2} = \frac{14}{14} = 1$, so these fractions are reciprocals.

Multiplication on the set of rational numbers has all of the same properties as multiplication on the set of whole numbers, and one new property, the **inverse property of multiplication**. Here is the complete list of properties of multiplication:

- The **commutative property of multiplication**: if r and s are any two rational numbers, $r \cdot s = s \cdot r$.

- The **associative property of multiplication**: if r, s, and t are any three rational numbers, $r \cdot (s \cdot t) = (r \cdot s) \cdot t$.

- The **identity property of multiplication**: if r is any rational number, $1 \cdot r = r \cdot 1 = r$. The number 1 is called the **multiplicative identity**.

- The **inverse property of multiplication**: if r is any rational number other than 0, there exists a unique rational number s such that $r \cdot s = s \cdot r = 1$. Such rational numbers are **multiplicative inverses**, or **reciprocals**, of each other.

- The **zero product property of multiplication**: if r is any rational number, $r \cdot 0 = 0 \cdot r = 0$, and if $r \cdot s = 0$ then at least one of the factors, r or s, must be 0.

Practice B

Test whether the following pairs of fractions are reciprocals. Turn the page to check your solutions.

7. $\frac{2}{3}$ and $\frac{3}{2}$

8. $\frac{-2}{3}$ and $\frac{3}{-2}$

9. $\frac{2}{14}$ and $\frac{7}{4}$

10. $\frac{3}{9}$ and $\frac{18}{6}$

11. $\frac{7}{4}$ and $-\frac{4}{7}$

12. $-\frac{4}{6}$ and $-\frac{12}{8}$

C. No More Division!

In Chapter 2, we introduced the **addition property of equations**: When you add the same number to both sides of an equation (to the left and right of the equals sign), the resulting equation has exactly the same solution or solutions as the original equation. There is a similar property for multiplication:

Multiplication Property of Equations

When you multiply both sides of an equation by the same non-zero number, the resulting equation has exactly the same solution(s) as the original equation.

These properties of addition and multiplication are at the core of the algebra-centered approach to solving equations that we are developing in this book. We'll use the multiplication property of equations now to help us find a new approach to rational number division.

The definition of division for rational numbers is about the same as with whole numbers and integers. Now we can evaluate quotients of any integers as long as the divisor is not zero!

Division

Let r and s be any two rational numbers with $s \neq 0$. The *quotient* of r and s is the unique rational number q such that $r = q \cdot s$. In other words, the equations $r \div s = q$ and $r = q \cdot s$ are *equivalent*.

According to the multiplication property of equations, we can multiply both sides of $r = q \cdot s$ by any non-zero number, and the resulting equation will be equivalent to $r = q \cdot s$. Let's use the symbol s' to represent the reciprocal of s. We know that $s' \neq 0$, so by the multiplication property of equations, the equations $r = q \cdot s$ and $r \cdot s' = q \cdot s \cdot s'$ are equivalent. We use properties of multiplication to simplify the right-hand side of this equation:

$$q \cdot s \cdot s' = q \cdot (s \cdot s') \qquad \text{Apply the associative property.}$$
$$= q \cdot 1 \qquad \text{Apply the inverse property, because } s' \text{ is the reciprocal of } s.$$
$$= q \qquad \text{By the identity property.}$$

Finally, in the following sequence of equations, each is equivalent to the next:

$$r \div s = q$$
$$r = q \cdot s$$
$$r \cdot s' = q \cdot s \cdot s'$$
$$r \cdot s' = q$$

The variable q represents the *quotient* of r and s. The last equation on this list can be stated like this: "The **product** of r and the **reciprocal** of s is equal to the **quotient** of r and s."

This statement has a powerful implication. In the set of rational numbers, we no longer need the operation of division! Every division can be rewritten as an **equivalent multiplication**.

Practice A — Answers

1. $\dfrac{10}{21}$

2. $\dfrac{2}{5}$

3. $\dfrac{3}{2}$

4. $\dfrac{26}{7}$

5. $-\dfrac{5}{26}$

6. 1

Equivalent Multiplication

Let r and s be any two rational numbers with $s \neq 0$ and let s' be the reciprocal of s. If $r \div s = q$, then $r \cdot s' = q$ as long as $s \neq 0$.

Example 3

Rewrite each quotient of rational numbers as an equivalent product. Then evaluate and reduce the product.

1. $\dfrac{12}{5} \div \dfrac{36}{25}$

2. $\dfrac{7/16}{35/8}$

Solutions

1.

$$\frac{12}{5} \div \frac{36}{25} = \frac{12}{5} \cdot \frac{25}{36}$$

The equivalent multiplication is the product of the first fraction and the reciprocal of the second fraction.

$$= \frac{2 \cdot 2 \cdot 3 \cdot 5 \cdot 5}{5 \cdot 2 \cdot 2 \cdot 3 \cdot 3}$$

Factor the numerator and denominator until all numbers are prime (prime factorization).

$$= \frac{2 \cdot 2 \cdot 3 \cdot 5 \cdot 5}{2 \cdot 2 \cdot 3 \cdot 3 \cdot 5}$$

Apply the associative property and commutative property until the factors of the numerator and denominator are in increasing order.

$$= 1 \cdot 1 \cdot 1 \cdot 1 \cdot \frac{5}{3}$$

Since $\frac{n}{n} = 1$, rewrite equivalent numbers.

$$= \frac{5}{3}$$

Apply the identity property.

2.

$$\frac{7/16}{35/8} = \frac{7}{16} \div \frac{35}{8}$$

Rewrite so the fractions are easier to work with.

$$= \frac{7}{16} \cdot \frac{8}{35}$$

The equivalent multiplication is the product of the first fraction and the reciprocal of the second fraction.

$$= \frac{7 \cdot 2 \cdot 2 \cdot 2}{2 \cdot 2 \cdot 2 \cdot 2 \cdot 7 \cdot 5}$$

Factor the numerator and denominator until all numbers are prime (prime factorization).

$$= \frac{2 \cdot 2 \cdot 2 \cdot 7}{2 \cdot 2 \cdot 2 \cdot 2 \cdot 5 \cdot 7}$$

Apply the associative property and commutative property until the factors of the numerator and denominator are in increasing order.

$$= 1 \cdot 1 \cdot 1 \cdot 1 \cdot \frac{1}{2 \cdot 5}$$

Since $\frac{n}{n} = 1$, rewrite equivalent numbers.

$$= \frac{1}{10}$$

Apply the identity property and evaluate the denominator.

Practice C

Rewrite each quotient of rational numbers as an equivalent product. Then evaluate and reduce the product. Turn the page to check your solutions.

13. $\frac{2}{3} \div \frac{4}{9}$

14. $\frac{1/3}{2/3}$

15. $\frac{3/11}{9/7}$

16. $\frac{3}{4} \div \frac{15}{8}$

17. $\frac{-\frac{5}{9}}{\frac{7}{12}}$

18. $\frac{4/12}{10/16}$

Exercises 3.5

For the following exercises, evaluate and reduce each product.

1. $\frac{2}{3} \cdot \frac{5}{8}$

2. $\frac{4}{3} \cdot \frac{9}{5}$

3. $\frac{-2}{3} \cdot \frac{9}{-8}$

4. $\frac{1}{5} \cdot \frac{15}{4}$

5. $-\frac{2}{3} \cdot \frac{5}{7}$

6. $\frac{4}{3} \cdot \frac{-5}{9}$

7. $\frac{3}{-2} \cdot \frac{9}{-8}$

8. $-\frac{2}{5} \cdot \left(-\frac{4}{3}\right)$

9. $\frac{4}{3} \cdot \frac{9}{10}$

10. $\frac{-8}{11} \cdot \frac{5}{-4}$

11. $\frac{4}{3} \cdot \left(-\frac{9}{5}\right)$

12. $-\frac{15}{2} \cdot \frac{7}{6}$

13. $-\frac{12}{25} \cdot \frac{10}{3}$

14. $\frac{6}{35} \cdot \left(-\frac{14}{15}\right)$

15. $\frac{99}{67} \cdot \frac{134}{55}$

16. $\frac{5}{-6} \cdot \frac{-8}{5}$

17. $\frac{5}{-7} \cdot \frac{-4}{11}$

18. $\frac{12}{28} \cdot \frac{35}{15}$

19. $\frac{1}{33} \cdot \frac{3}{5}$

20. $-\frac{44}{24} \cdot \left(-\frac{20}{16}\right)$

21. $-\frac{30}{24} \cdot \left(-\frac{66}{55}\right)$

22. $\frac{7}{23} \cdot \frac{0}{71}$

23. $-\frac{79}{32} \cdot \frac{2}{395}$

24. $\frac{33}{9} \cdot \left(-\frac{0}{14}\right)$

25. $\frac{0}{34} \cdot \left(-\frac{29}{18}\right)$

26. $-\frac{36}{27} \cdot \frac{9}{12}$

For the following exercises, test each pair of fractions to find out if they are reciprocals.

27. $\frac{4}{9}$ and $\frac{9}{4}$

28. $-\frac{5}{13}$ and $-\frac{13}{5}$

29. $\frac{0}{7}$ and $\frac{7}{0}$

30. $\frac{24}{40}$ and $\frac{25}{15}$

31. $\frac{6}{14}$ and $\frac{16}{4}$

32. $\frac{26}{39}$ and $-\frac{15}{10}$

33. $\frac{4}{1}$ and $\frac{1}{4}$

34. $\frac{7}{1}$ and $\frac{2}{14}$

35. $-\frac{4}{6}$ and $-\frac{18}{12}$

36. $\frac{2}{4}$ and $\frac{8}{16}$

37. $\frac{2}{3}$ and $-\frac{3}{2}$

38. $\frac{1}{9}$ and $\frac{9}{1}$

For the following exercises, rewrite each quotient of rational numbers as an equivalent product, and then evaluate and reduce the product.

39. $\frac{3}{4} \div \frac{9}{2}$

40. $\frac{5/8}{15/16}$

41. $\frac{3/2}{1/3}$

42. $\frac{2/5}{8/5}$

43. $\frac{7/4}{5/3}$

44. $\frac{2/3}{5/9}$

45. $\frac{4/9}{6/15}$

46. $\frac{1/3}{6/5}$

47. $\frac{0/13}{5/6}$

48. $\frac{0/5}{12/7}$

49. $\frac{3}{7} \div \frac{9}{4}$

50. $\frac{-34/3}{-4/6}$

51. $\frac{14/9}{-21/6}$

52. $-\frac{8}{15} \div \frac{4}{25}$

Use either multiplication or division to answer each question.

53. Monica is able to type one page in $\frac{5}{12}$ of an hour. Her research paper is 25 pages long. How many hours will it take Monica to finish typing the paper?

54. A garden box is $\frac{5}{3}$ yards wide and $\frac{11}{4}$ yards long. How many square yards of area does the garden box cover?

55. The instructions on a package of wildflower seed mix reads "scatter the seeds evenly over $\frac{7}{4}$ square yards." Bryce wants to plant $\frac{21}{8}$ square yards of wildflowers. How many packages of seed will he need?

56. Mindy completes a road race in $\frac{7}{8}$ of an hour. If she runs each mile in $\frac{2}{15}$ of an hour, how many miles long is the race?

57. James used 24 pounds of gravel split evenly to supply his fish tanks. Each tank received $\frac{3}{2}$ pounds of gravel. How many fish tanks did he supply?

58. Julie used $\frac{2}{3}$ pounds of gravel for each fish tank. She supplied 14 fish tanks with gravel. How much gravel did she use, total?

59. A group of students evenly split $\frac{4}{3}$ of a pizza. They each get $\frac{2}{9}$ of a whole pizza. How many students ate pizza?

60. If four people divide $\frac{2}{3}$ of a pizza evenly between them, how much of the whole pizza does each person get?

61. The age of a certain deer is two thirds of the age of a certain badger. The badger is four years old. Find the age of the deer.

62. The age of the cow is three fifths of the age of the horse. The cow is seven years old. Find the horse's age.

13. $\frac{2}{3} \div \frac{4}{9} = \frac{3}{2}$

14. $\frac{1/3}{2/3} = \frac{1}{2}$

15. $\frac{3/11}{9/7} = \frac{7}{33}$

16. $\frac{3}{4} \div \frac{15}{8} = \frac{2}{5}$

17. $\frac{-\frac{5}{9}}{\frac{7}{12}} = -\frac{20}{21}$

18. $\frac{4/12}{10/16} = \frac{8}{15}$

3.6 Comparing Rational Numbers

In Section 3.4, you learned how to determine if two fractions represent the same rational number. In this section, you'll learn how to determine whether one rational number in fraction notation is less than or greater than another rational number in fraction notation.

The reasoning that we use here is similar to the reasoning that we would use to help Allen and Raquel find a way to answer the following questions:

1. Allen has 8 apples, and Raquel has 5 apples. Who has more fruit?

2. Allen has 8 watermelons, and Raquel has 5 grapes. Who has more fruit?

3. Allen has 2 watermelons, and Raquel has 900 grapes. Who has more fruit?

A. Fractions with the Same Positive Denominator

In question 1 about apples, we can make a direct comparison between the numbers of fruit because each piece of fruit is the same. Eight of one thing is more than five of the same thing because eight is greater than five. We can use this type of reasoning to compare fractions with the same positive denominator:

$$\text{If } a > b \text{ and } c > 0 \text{ then } \frac{a}{c} > \frac{b}{c}$$

Example 1

Compare the following fractions by filling in the blank with > or <.

1. $\frac{8}{29}$ —— $\frac{5}{29}$

2. $\frac{-8}{29}$ —— $\frac{5}{29}$

3. $\frac{-8}{-29}$ —— $\frac{5}{-29}$

Solutions

1. $\frac{8}{29} > \frac{5}{29}$ because 29 is positive and $8 > 5$. This is the same reasoning that we use to determine that Allen's 8 apples are more fruit than Raquel's 5 apples.

2. $\frac{-8}{29} < \frac{5}{29}$ because 29 is positive and $-8 < 5$. This rule still applies regardless of whether the numerators are positive or negative.

3. We can't use the rule immediately in this example because the denominators are not positive. We can rewrite these fractions as equivalent fractions with positive denominators and then compare those fractions: $\frac{-8}{-29} > \frac{5}{-29}$ because $\frac{-8}{-29} = \frac{8}{29}, \frac{5}{-29} = \frac{-5}{29}$, and $\frac{8}{29} > \frac{-5}{29}$.

Practice A

Compare each pair of fractions by filling in the blank with > or <, whichever creates a true statement. Turn the page to check your solutions.

1. $\dfrac{23}{17}$ ___ $\dfrac{32}{17}$

2. $\dfrac{-4}{9}$ ___ $\dfrac{-7}{9}$

3. $\dfrac{4}{-9}$ ___ $\dfrac{7}{-9}$

4. $\dfrac{215}{365}$ ___ $\dfrac{200}{365}$

5. $\dfrac{-5}{7}$ ___ $\dfrac{-6}{7}$

6. $\dfrac{6}{-5}$ ___ $\dfrac{1}{-5}$

B. Fractions with Different Denominators

Let's go back to Allen and Raquel and their battle of the fruit. With questions 2 and 3, Allen has watermelons, and Raquel has grapes. We want to know who has more *fruit*, not just who has more watermelon, apples, or grapes.

In these questions, it doesn't make sense to directly compare the number of pieces of fruit. You could guess that 8 watermelons must be more than 5 grapes in question 2 because a watermelon is bigger than a grape and there are more watermelons than grapes. You'd probably be right. There are some situations where you can easily compare fractions, even if they have different denominators. Below are three such situations.

First, if one fraction represents a negative rational number and the other represents a positive rational number, the positive rational number is always greater. Remember that every negative number is automatically less than any positive number.

Second, if both fractions represent positive rational numbers, but one is less than 1 and the other is greater than 1, then the second rational number is greater. In a fraction with a positive numerator and denominator, the denominator is the number of portions that make up one whole. If the numerator (the number of portions), is greater than the denominator (the number of portions that make up one whole), the fraction represents a rational number greater than 1. A fraction like this is called an **improper fraction**.

On the other hand, if the numerator is less than the denominator, then the fraction represents a rational number that is less than one. A fraction like this is called a **proper fraction**. Every positive proper fraction is always less than any positive improper fraction.

Combining these first two facts gives us the following comparison of positive or negative fractions that are either improper or proper (Figure 1):

A third situation where it's easy to compare fractions is when the numerators and denominators of both fractions are positive, but one fraction has a numerator that is **less than** the numerator of the other fraction and a denominator

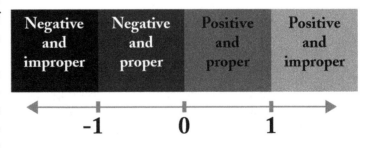

Figure 1

which is **greater than** the denominator of the other fraction. This fraction represents fewer and smaller portions than the other fraction. In the same way that 5 grapes are clearly less fruit than 8 watermelons, the fraction with a lesser numerator and greater denominator is always less than the other fraction.

Example 2

Compare the following fractions by filling in the blank with > or <.

1. $\frac{5}{29}$ ___ $\frac{8}{23}$
2. $\frac{-5}{4}$ ___ $\frac{6}{17}$
3. $\frac{3}{2}$ ___ $\frac{5}{7}$

Solution

1. $\frac{5}{29}$ < $\frac{8}{23}$ The first fraction represents fewer portions than the second fraction, and the first fraction represents smaller portions than the second fraction. This is like comparing five grapes to eight watermelons.

2. $\frac{-5}{4}$ < $\frac{6}{17}$ The first fraction is negative, and the second is positive.

3. $\frac{3}{2}$ > $\frac{5}{7}$ The second fraction is proper, and the first fraction is improper.

Practice B

Compare each pair of fractions by filling in the blank with > or <, whichever creates a true statement. Turn the page to check your solutions.

7. $\frac{1}{2}$ ___ $\frac{5}{3}$

8. $-\frac{1}{2}$ ___ $-\frac{5}{3}$

9. $\frac{7}{2}$ ___ $\frac{5}{3}$

10. $\frac{-5}{21}$ ___ $\frac{2}{-3}$

11. $\frac{5}{4}$ ___ $\frac{6}{17}$

12. $\frac{3}{8}$ ___ $\frac{5}{7}$

C. Finding a Common Denominator

Let's go back to Allen and Raquel and their fruit again. The correct answer to question 3 is far from clear. Allen has 2 watermelons, and Raquel has 900 grapes. Who has more fruit? It's hard to say! How can we judge accurately whether Allen's 2 watermelons are more fruit or less fruit than Raquel's 900 grapes?

In the real world, we would probably weigh the fruit so that we could make a direct comparison of their weight. To compare fractions with different denominators, we have to do the same sort of thing. We have to rewrite the given fractions as equivalent fractions which have equal positive denominators.

Think about pizza as an example. In an eating contest, Jan ate $\frac{5}{8}$ of a pizza while Cole ate $\frac{7}{12}$ of an identical pizza. Who ate more pizza? This is like comparing watermelons and grapes. Cole ate more *slices* than Jan, but Jan's slices were bigger than Cole's slices because Jan's pizza was cut into fewer slices.

In Figure 2, we see Jan's pizza divided into eight pieces on the left. The five that Jan ate are shaded. Now imagine every slice of Jan's pizza was divided into 3 equal pieces along the dotted lines.

The entire pizza is then cut into 24 equal slices on the right, and we see that Jan ate 15 of them. Because the fractions $\frac{5}{8}$ and $\frac{15}{24}$ are equivalent, it's also correct to say that Jan ate $\frac{15}{24}$ of a pizza.

In Figure 3, the 7 pieces that Cole ate are shaded on the left. If we divide every slice of Cole's pizza into 2 equal pieces, the entire pizza will also be cut into 24 equal slices on the right.

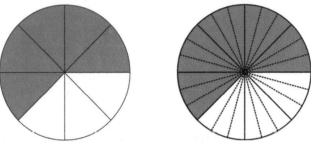

Figure 2. Jan's Pizza

Because the fractions $\frac{7}{12}$ and $\frac{14}{24}$ are equivalent, it's also correct to say that Cole ate $\frac{14}{24}$ of a pizza.

Using the equivalent fractions with common denominators, we can represent Jan's accomplishment as $\frac{15}{24}$ of a pizza and Cole's as $\frac{14}{24}$ of a pizza instead of the given fractions, $\frac{5}{8}$ and $\frac{7}{12}$. We can now see that

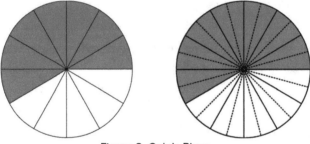

Figure 3. Cole's Pizza

Jan won the pizza-eating contest. Here is our work:

$$\frac{5}{8} = \frac{5 \cdot 3}{8 \cdot 3} = \frac{15}{24}$$

$$\frac{7}{12} = \frac{7 \cdot 2}{12 \cdot 2} = \frac{14}{24}$$

The results: $\frac{15}{24} > \frac{14}{24}$, so $\frac{5}{8} > \frac{7}{12}$

How do we know to represent these rational numbers with fractions that have 24 as the denominator? The key is that 24 is the **least common multiple** of the given denominators 8 and 12.

These are the positive multiples of 8:

$$8, 16, 24, 32, 40, \ldots$$

These are the positive multiples of 12:

$$12, 24, 36, 48, 60, \ldots$$

We can see from these lists that LCM(8, 12) = 24, so we rewrite $\frac{5}{8}$ and $\frac{7}{12}$ as equivalent fractions with denominators of 24.

Example 3

Compare the fractions $\frac{11}{15}$ and $\frac{16}{21}$.

Solution

These fractions have different denominators and cannot be easily compared, so we have to rewrite them as equivalent fractions with the same denominator. That denominator should be LCM(15, 21) = 105, as shown below.

Multiples of 15: 15, 30, 45, 60, 75, 90, 105, 120, …

Multiples of 21: 21, 42, 63, 84, 105, 126, 147, 168, …

$$\frac{11}{15} = \frac{11 \cdot 7}{15 \cdot 7} = \frac{77}{105} \text{ and } \frac{16}{21} = \frac{16 \cdot 5}{21 \cdot 5} = \frac{80}{105}$$

On inspection, we see that $\frac{77}{105} < \frac{80}{105}$, so $\frac{11}{15} < \frac{16}{21}$.

Practice C

Compare each pair of fractions by filling in the blank with > or <, whichever creates a true statement. Turn the page to check your solutions.

13. $\frac{1}{3}$ _____ $\frac{3}{10}$

14. $\frac{3}{4}$ _____ $\frac{4}{5}$

15. $\frac{4}{3}$ _____ $\frac{19}{15}$

16. $\frac{5}{4}$ _____ $\frac{7}{6}$

17. $\frac{19}{24}$ _____ $\frac{7}{9}$

18. $\frac{23}{27}$ _____ $\frac{13}{15}$

Practice A — Answers

1. $\frac{23}{17} < \frac{32}{17}$

2. $\frac{-4}{9} > \frac{-7}{9}$

3. $\frac{4}{-9} > \frac{7}{-9}$

4. $\frac{215}{365} > \frac{200}{365}$

5. $\frac{-5}{7} > \frac{-6}{7}$

6. $\frac{6}{-5} < \frac{1}{-5}$

Practice B — Answers

7. $\frac{1}{2} < \frac{5}{3}$

8. $-\frac{1}{2} > -\frac{5}{3}$

9. $\frac{7}{2} > \frac{5}{3}$

10. $-\frac{5}{21} > \frac{2}{-3}$

11. $\frac{5}{4} > \frac{26}{17}$

12. $\frac{3}{8} < \frac{5}{7}$

Exercises 3.6

For the following exercises, use the corresponding letter below to indicate the method that you would use to compare each pair of fractions. Then fill in the blank with > or <, whichever creates a true statement.

       a. Compare numerators because denominators are equal and positive.
       b. Compare the sign because one is positive, and one is negative.
       c. Compare by whether each fraction is proper or improper.
       d. Compare number and size of portions.
       e. Rewrite as equivalent factions with the same positive denominator.

1. $\dfrac{2}{9}$ —— $\dfrac{4}{7}$

2. $\dfrac{2}{3}$ —— $\dfrac{4}{7}$

3. $\dfrac{-2}{9}$ —— $\dfrac{1}{7}$

4. $\dfrac{9}{8}$ —— $\dfrac{4}{7}$

5. $\dfrac{7}{9}$ —— $\dfrac{4}{9}$

6. $\dfrac{9}{7}$ —— $\dfrac{4}{7}$

7. $\dfrac{-9}{8}$ —— $\dfrac{-4}{7}$

8. $\dfrac{-2}{9}$ —— $\dfrac{-5}{9}$

9. $-\dfrac{12}{5}$ —— $-\dfrac{17}{7}$

10. $-\dfrac{8}{9}$ —— $\dfrac{4}{5}$

11. $\dfrac{8}{9}$ —— $\dfrac{14}{11}$

12. $-\dfrac{27}{7}$ —— $-\dfrac{9}{2}$

13. $\dfrac{5}{8}$ —— $\dfrac{4}{7}$

14. $-\dfrac{5}{8}$ —— $-\dfrac{14}{9}$

15. $-\dfrac{3}{2}$ —— $-\dfrac{5}{2}$

16. $\dfrac{5}{23}$ —— $\dfrac{7}{23}$

17. $-\dfrac{4}{7}$ —— $-\dfrac{5}{6}$

18. $-\dfrac{3}{4}$ —— $-\dfrac{2}{7}$

19. $-\dfrac{8}{7}$ —— $-\dfrac{9}{5}$

20. $\dfrac{3}{-14}$ —— $\dfrac{2}{-14}$

21. $\dfrac{7}{-9}$ —— $\dfrac{5}{-9}$

22. $\dfrac{4}{7}$ —— $\dfrac{5}{6}$

For the following exercises, rewrite each pair of fractions as equivalent fractions with the same positive denominators. Then fill in the blank with either > or <, whichever creates a true statement.

23. $\frac{3}{4}$ ___ $\frac{5}{8}$

24. $\frac{3}{4}$ ___ $\frac{5}{7}$

25. $\frac{3}{4}$ ___ $\frac{9}{11}$

26. $\frac{3}{4}$ ___ $\frac{10}{13}$

27. $\frac{4}{3}$ ___ $\frac{7}{5}$

28. $\frac{4}{3}$ ___ $\frac{8}{5}$

29. $\frac{7}{3}$ ___ $\frac{5}{4}$

30. $\frac{8}{3}$ ___ $\frac{5}{4}$

31. $\frac{8}{5}$ ___ $\frac{11}{6}$

32. $\frac{8}{5}$ ___ $\frac{5}{4}$

33. $\frac{11}{15}$ ___ $\frac{7}{9}$

34. $\frac{9}{16}$ ___ $\frac{7}{12}$

35. $\frac{11}{6}$ ___ $\frac{7}{9}$

36. $\frac{5}{4}$ ___ $\frac{7}{6}$

37. $\frac{11}{6}$ ___ $\frac{16}{9}$

38. $\frac{9}{8}$ ___ $\frac{13}{12}$

39. $\frac{-3}{4}$ ___ $\frac{5}{-7}$

40. $-\frac{3}{4}$ ___ $-\frac{5}{8}$

41. $-\frac{4}{3}$ ___ $-\frac{8}{5}$

42. $\frac{-4}{3}$ ___ $\frac{7}{-5}$

43. $\frac{-11}{-5}$ ___ $\frac{-19}{-8}$

44. $-\frac{14}{9}$ ___ $\frac{-19}{12}$

45. $\frac{9}{16}$ ___ $\frac{-7}{12}$

46. $\frac{-11}{-15}$ ___ $-\frac{7}{9}$

For the following exercises, write each list of fractions in order from least to greatest.

47. $\frac{13}{17}, \frac{4}{3},$ and $\frac{-2}{3}$

48. $\frac{5}{7}, \frac{3}{7},$ and $\frac{8}{7}$

49. $\frac{8}{5}, \frac{5}{8},$ and $\frac{4}{9}$

50. $\frac{5}{4}, \frac{8}{7},$ and $\frac{3}{7}$

51. $\frac{13}{17}, \frac{4}{3},$ and $\frac{5}{2}$

52. $\frac{3}{7}, \frac{8}{7},$ and $-\frac{4}{7}$

53. $\frac{2}{3}, -\frac{11}{6},$ and $-\frac{3}{4}$

54. $\frac{3}{7}, \frac{7}{6},$ and $\frac{5}{7}$

55. $\frac{7}{11}, \frac{4}{3},$ and $\frac{2}{3}$

56. $-\frac{2}{3}, \frac{9}{14},$ and $-\frac{4}{7}$

57. $-\frac{3}{7}, -\frac{5}{13},$ and $-\frac{6}{13}$

58. $\frac{8}{7}, \frac{3}{4}, \frac{5}{6},$ and $\frac{3}{7}$

59. $\frac{4}{9}, \frac{5}{3}, \frac{6}{7},$ and $\frac{7}{4}$

60. $\frac{5}{4}, \frac{7}{6}, \frac{5}{6},$ and $\frac{8}{7}$

61. $-\frac{4}{9}, \frac{5}{3}, \frac{6}{7},$ and $-\frac{7}{4}$

62. $\frac{2}{3}, \frac{3}{4}, \frac{5}{7},$ and $\frac{5}{6}$

3.7 Addition and Subtraction

In Section 3.6, we asked some pretty profound questions about whether Allen or Raquel had more fruit. The point of those questions was to show how rational numbers can be compared if they are represented as fractions. Now consider the following three new questions:

1. Allen has 8 apples, and Raquel has 5 apples. How much fruit do they have together?

2. Allen has 11 watermelons, and Raquel has 2 grapes. How much fruit do they have together?

3. Allen has 2 watermelons, and Raquel has 11 grapes. How much fruit do they have together?

A. Adding and Subtracting Fractions with the Same Denominator

Of these three questions, only the first can be answered in a reasonable way. Together, Allen and Raquel have 13 apples. This is not a special property of apples, either. The same goes for other objects as well:

$$8 \text{ canoes} + 5 \text{ canoes} = 13 \text{ canoes}$$

$$8 \text{ bowties} + 5 \text{ bowties} = 13 \text{ bowties}$$

$$8 \text{ beavers} + 5 \text{ beavers} = 13 \text{ beavers}$$

$$8 \text{ shrubs} + 5 \text{ shrubs} = 13 \text{ shrubs}$$

$$8 \text{ sevenths} + 5 \text{ sevenths} = 13 \text{ sevenths}$$

The last example on the list is actually a sum of fractions, and it fits the pattern of all of the previous examples. When we add fractions that have the same denominator, we simply add the numerators and use the same denominator that the fractions have.

> ### Adding Fractions with the Same Denominator
>
> If a, b, and c are integers and $c \neq 0$, then $\frac{a}{c} + \frac{b}{c} = \frac{a+b}{c}$.

Example 1

Evaluate each sum. Reduce your answer if possible.

1. $\frac{8}{7} + \frac{5}{7}$

2. $\frac{8}{5} + \frac{-3}{5}$

Solutions

1. $\frac{8}{7} + \frac{5}{7} = \frac{13}{7}$, Add together the two numerators and use the same denominator. The fraction can't be reduced.

2. $\frac{8}{5} + \frac{-3}{5} = \frac{5}{5} = 1$, Add together the two numerators and use the same denominator. The fraction can be reduced because the numerator and denominator are the same.

Rational number addition has all of the same properties as integer addition. Here is a complete list of the properties of addition.

- The **commutative property of addition**: when r and s are any two rational numbers, then $r + s = s + r$

- The **associative property of addition**: when r, s, and t are any three rational numbers, then $(r + s) + t = r + (s + t)$

- The **identity property of addition**: when r is any rational number, then $r + 0 = 0 + r = r$

- The **inverse property of addition**: every rational number r has an additive inverse $-r$, such that $r + (-r) = -r + r = 0$

Subtracting a rational number r is the same as adding $-r$. We can rewrite differences of rational numbers as equivalent sums in exactly the same way that we can for integers.

Example 2

Rewrite the following differences as equivalent sums, then evaluate the sums. Reduce your answers if possible.

1. $\frac{3}{5} - \frac{2}{5}$ **2.** $\frac{7}{13} - \left(-\frac{5}{13}\right)$ **3.** $-\frac{4}{9} - \frac{11}{9}$

Solutions

1. $\frac{3}{5} - \frac{2}{5} = \frac{3}{5} + \left(-\frac{2}{5}\right) = \frac{1}{5}$

2. $\frac{7}{13} - \left(-\frac{5}{13}\right) = \frac{7}{13} + \frac{5}{13} = \frac{12}{13}$

3. $-\frac{4}{9} - \frac{11}{9} = -\frac{4}{9} + \left(-\frac{11}{9}\right) = -\frac{15}{9} = -\frac{5}{3}$

Note: sometimes the negative symbol is in the numerator or the denominator, and other times it's in line with the fraction bar. The expressions $\frac{-a}{b}$, $\frac{a}{-b}$, and $-\frac{a}{b}$ are all equivalent to each other.

Practice A

Evaluate each sum or difference, then reduce if possible. Turn the page to check your solutions.

1. $\frac{3}{4} + \frac{7}{4}$

2. $\frac{-7}{6} + \frac{7}{6}$

3. $\frac{25}{12} - \left(\frac{-17}{12}\right)$

4. $\frac{-5}{8} + \frac{-9}{8}$

5. $\frac{13}{7} - \frac{8}{7}$

6. $-\frac{6}{15} - \left(-\frac{11}{15}\right)$

B. Adding Fractions with Different Denominators

After bonding over their mutual obsession with fruit, Allen and Raquel have fallen in love and decided to share their fruit. The apparent answer to questions 2 and 3 is the same: Together Allen and Raquel have 13 pieces of fruit. But this doesn't make very much sense! No one who has ever shopped for watermelons and grapes would accept the proposition that 11 watermelons and 2 grapes is the same amount of fruit as 2 watermelons and 11 grapes.

In the same way, if two rational numbers are represented by fractions with different denominators, then we can't use those fractions to evaluate the sum or difference of the rational numbers. We must first rewrite the fractions as equivalent fractions that have the same denominator. As we saw in Section 3.6, we can use the least common multiple of the denominators as our common denominator.

Example 3

Evaluate each sum or difference. Reduce if possible.

1. $\frac{11}{3} + \frac{2}{15}$

2. $\frac{11}{15} + \frac{2}{3}$

3. $\frac{5}{8} - \frac{7}{12}$

Solutions

1. Because 15 is already a multiple of 3, we only have to rewrite one fraction

$$\frac{11}{3} + \frac{2}{15} = \frac{11 \cdot 5}{3 \cdot 5} + \frac{2}{15}$$

$$= \frac{55}{15} + \frac{2}{15}$$

$$= \frac{57}{15}$$

$$= \frac{19}{5}$$

2. Like the previous example, we only have to rewrite one fraction:

$$\frac{11}{15} + \frac{2}{3} = \frac{11}{15} + \frac{2 \cdot 5}{3 \cdot 5}$$

$$= \frac{11}{15} + \frac{10}{15}$$

$$= \frac{21}{15}$$

$$= \frac{7}{5}$$

3. In this problem, the least common multiple of the denominators is 24:

$$\frac{5}{8} - \frac{7}{12} = \frac{5 \cdot 3}{8 \cdot 3} - \frac{7 \cdot 2}{12 \cdot 2}$$

$$= \frac{15}{24} - \frac{14}{24}$$

$$= \frac{1}{24}$$

The first two problems above are inspired by Allen and Raquel. One-third portions are greater than one-fifteenth portions. Think of a one-third portion like a watermelon, while a one-fifteenth portion is more like a grape. In the same way that 11 watermelons and 2 grapes is more fruit than 11 grapes and 2 watermelons, the first two examples show us that $\left(\frac{11}{3} + \frac{2}{15}\right) > \left(\frac{11}{15} + \frac{2}{3}\right)$ because $\frac{57}{15} > \frac{21}{15}$.

The third problem above is inspired by Jan and Cole's pizza-eating contest from Section 3.6. In that contest, you learned that Jan ate more pizza than Cole because $\frac{5}{8} > \frac{7}{12}$. However, the analysis doesn't tell us the margin of victory. What is the difference of $\frac{5}{8}$ and $\frac{7}{12}$? In problem 3 in the example above, we learn that Jan won by $\frac{1}{24}$ of a pizza because $\frac{5}{8} - \frac{7}{12} = \frac{1}{24}$. It turns out that it was a very close contest. Cole has nothing to be ashamed of.

Practice B

Evaluate each sum or difference. Reduce your answer if possible. Turn the page to check your solutions.

7. $\frac{4}{3} + \frac{5}{6}$

8. $\frac{1}{2} - \frac{2}{3}$

9. $\frac{5}{18} - \frac{5}{6}$

10. $\frac{5}{6} + \frac{2}{15}$

11. $-\frac{7}{18} - \frac{5}{12}$

12. $\frac{7}{15} - \frac{4}{9} + \frac{3}{10}$

Exercises 3.7

For the following exercises, evaluate each sum or difference. Reduce your answer if possible.

1. $\frac{3}{5} + \frac{4}{5}$

2. $\frac{2}{5} + \frac{8}{5}$

3. $-\frac{2}{3} - \left(-\frac{8}{3}\right)$

4. $\frac{7}{15} - \left(-\frac{4}{15}\right)$

5. $\frac{4}{9} - \frac{7}{9}$

6. $-\frac{6}{11} - \frac{5}{11}$

7. $-\frac{25}{14} - \frac{10}{14}$

8. $-\left(-\frac{1}{3}\right) - \left(-\frac{1}{3}\right)$

9. $-\left(-\frac{2}{3}\right) - \frac{4}{3}$

10. $1 + \frac{1}{2}$

11. $1 + \frac{1}{3}$

12. $1 + \frac{6}{5}$

13. $1 + \frac{3}{8}$

14. $3 - \frac{2}{7}$

15. $\frac{7}{4} + 5$

16. $\frac{2}{9} - 2$

17. $\frac{5}{2} - \frac{3}{4}$

18. $\frac{2}{5} - \frac{1}{15}$

19. $\frac{2}{3} - \frac{4}{9}$

20. $\frac{1}{4} + \frac{5}{12}$

21. $-\frac{4}{7} - \left(-\frac{24}{35}\right)$

22. $\frac{13}{24} + \left(-\frac{3}{8}\right)$

23. $\frac{3}{8} + \frac{1}{2}$

24. $\frac{1}{6} + \frac{1}{8}$

25. $\frac{2}{3} + \frac{1}{4}$

26. $\frac{3}{4} - \frac{2}{3}$

27. $-\frac{5}{12} + \frac{11}{18}$

28. $\frac{13}{20} - \left(-\frac{7}{15}\right)$

29. $-\left(-\frac{4}{3}\right) - \left(-\frac{1}{6}\right)$

30. $-\left(-\frac{4}{7}\right) - \frac{4}{7}$

31. $\frac{3}{8} + \frac{7}{12} - \frac{11}{15}$

Practice A — Answers

1. $\frac{3}{4} + \frac{7}{4} = \frac{10}{4} = \frac{5}{2}$

2. $\frac{-7}{6} + \frac{7}{6} = \frac{0}{6} = 0$

3. $\frac{25}{12} - \left(\frac{-17}{12}\right) = \frac{42}{12} = \frac{7}{2}$

4. $\frac{-5}{8} + \frac{-9}{8} = \frac{-14}{8} = -\frac{7}{4}$

5. $\frac{13}{7} - \frac{8}{7} = \frac{5}{7}$

6. $-\frac{6}{15} - \left(-\frac{11}{15}\right) = \frac{5}{15} = \frac{1}{3}$

32. $-\frac{5}{6} + \left(-\frac{3}{4}\right) - \left(-\frac{10}{7}\right)$

33. $-\frac{5}{4} + \frac{5}{4} - \left(-\frac{4}{5}\right)$

34. $\frac{3}{4} + \frac{1}{3} - \frac{2}{5} + 2$

35. $\frac{13}{18} - \left(-\frac{7}{20}\right) - \frac{8}{15}$

36. $1 + \frac{1}{2} + \frac{1}{3} + \frac{1}{4} + \frac{1}{5}$

37. $\frac{3}{4} - \frac{1}{4} \cdot \frac{5}{3}$

For the following exercises, use the correct order of operations to simplify each expression. Reduce your answer if possible.

38. $\frac{5}{7} - \frac{4}{9} \cdot \frac{6}{14}$

39. $\left(\frac{2}{3} - \frac{3}{4}\right) \cdot \frac{1}{2}$

40. $\frac{2}{3} - \frac{3}{4} \cdot \frac{1}{2}$

41. $\dfrac{\frac{4}{7} - \frac{1}{7}}{\frac{2}{49} + \frac{4}{49}}$

42. $\dfrac{\frac{3}{4} - \frac{1}{4}}{\frac{2}{5} + \frac{4}{5}}$

43. $\left(\frac{5}{8} - \frac{1}{12}\right) \cdot \frac{6}{7}$

44. $\dfrac{\frac{3}{4} - \frac{1}{3}}{\frac{2}{9} + \frac{1}{6}}$

45. $\dfrac{\left(\frac{2}{3} + \frac{1}{2}\right) \cdot 6}{8 \cdot \left(\frac{3}{4} - \frac{5}{2}\right)}$

46. $\dfrac{\frac{2}{3} + \frac{1}{2}}{\left(\frac{2}{3} - \frac{1}{2}\right) \cdot 2}$

47. $\frac{9}{16} \cdot \dfrac{\frac{1}{3} + \frac{1}{2}}{\frac{2}{3} - \frac{1}{4}}$

48. $\left(\frac{3}{4} - \frac{5}{12}\right) \cdot 6$

49. $\frac{3}{4} - \frac{5}{12} \cdot 6$

50. $\dfrac{2 - \frac{3}{4}}{\frac{2}{5} + 3}$

51. $\dfrac{3\left(2 + \frac{5}{6}\right)}{4\left(\frac{5}{2} + \frac{5}{4}\right)}$

52. $\dfrac{2 - \frac{3}{4}}{3}$

3.8 Solving Equations

Earlier in this book, we've used fact families to solve equations. At first, we operated with the set of whole numbers, which lacks both multiplicative and additive inverses. Then we moved on to the set of integers, which has additive inverses but lacks multiplicative inverses. Now that we are operating in the set of rational numbers, we no longer need fact families to solve equations. Instead, we can use the addition and multiplication properties of equations.

A. Using Addition and Multiplication Properties to Solve Equations

To refresh your memory, here are the addition and multiplication properties of equations:

Addition Property of Equations

If we add the same number to both sides of an equation, the resulting equation has exactly the same solutions as the original equation.

Multiplication Property of Equations

If we multiply both sides of an equation by the same non-zero number, the resulting equation has exactly the same solutions as the original equation.

In Chapter 2, we used the addition property to solve equations. The algebra for using the addition property has not changed, but the arithmetic is now a bit more complicated because we are adding rational numbers. Let's see how that works in the following example.

Example 1

Use the addition principle to solve each equation

1. $v + \frac{2}{3} = \frac{1}{5}$

2. $\frac{3}{4} - r = \frac{2}{3}$

Solutions

1. $v + \frac{2}{3} = \frac{1}{5}$

$\qquad v + \frac{2}{3} + \left(-\frac{2}{3}\right) = \frac{1}{5} + \left(-\frac{2}{3}\right)$ Add the opposite of $\frac{2}{3}$ to both sides.

$\qquad\qquad v + 0 = \frac{1}{5} + \left(-\frac{2}{3}\right)$ Apply the inverse property: $\frac{2}{3} + \left(-\frac{2}{3}\right) = 0$

$\qquad\qquad\qquad v = \frac{1}{5} + \left(-\frac{2}{3}\right)$ Apply the identity property: $v + 0 = v$

$\qquad\qquad\qquad v = \frac{3}{15} + \frac{-10}{15}$ Find LCM$(5, 3)$ and rewrite the fractions

$\qquad\qquad\qquad v = \frac{-7}{15}$ Evaluate the sum.

Our solution is $v = \frac{-7}{15}$. Now let's check our work.

$\qquad\qquad \frac{-7}{15} + \frac{2}{3} = \frac{1}{5}$ First we substitute our solution for v in the original equation.

$\qquad\qquad \frac{-7}{15} + \frac{10}{15} = \frac{1}{5}$ Find a common denominator.

$\qquad\qquad\qquad \frac{3}{15} = \frac{1}{5}$ Simplify by reducing the fraction.

$\qquad\qquad\qquad \frac{1 \cdot 3}{5 \cdot 3} = \frac{1}{5} \checkmark$ Our solution checks!

2. $\frac{3}{4} - r = \frac{2}{3}$

$\qquad\qquad \frac{3}{4} + (-r) = \frac{2}{3}$ Rewrite the subtraction as an equivalent addition.

$\qquad\qquad \frac{3}{4} + (-r) + r = \frac{2}{3} + r$ Add the opposite of $-r$ to both sides.

$\qquad\qquad \frac{3}{4} + 0 = \frac{2}{3} + r$ Apply the inverse property.

$\qquad\qquad \frac{3}{4} = \frac{2}{3} + r$ Apply the identity property.

$\qquad -\frac{2}{3} + \frac{3}{4} = -\frac{2}{3} + \frac{2}{3} + r$ Add the opposite of $\frac{2}{3}$ to both sides.

$\qquad -\frac{2}{3} + \frac{3}{4} = r$ Inverse and identity properties on the right hand side.

$\qquad -\frac{8}{12} + \frac{9}{12} = r$ Evaluate the sum.

$\qquad\qquad \frac{1}{12} = r$

Our solution is $\frac{1}{12} = r$. Let's check our work:

$$\frac{3}{4} - \frac{1}{12} = \frac{2}{3}$$

$$\frac{9}{12} - \frac{1}{12} = \frac{2}{3}$$

$$\frac{8}{12} = \frac{2}{3}$$

$$\frac{2 \cdot 4}{3 \cdot 4} = \frac{2}{3} \checkmark \qquad \text{Our solution checks!}$$

We can also use the multiplication principle to solve equations in which the variable is multiplied by a number. The key fact in each of these examples is that for any non-zero number a:

$$\frac{1}{a} \cdot ax = \left(\frac{1}{a} \cdot a\right) \cdot x = 1 \cdot x = x$$

To justify this, we use the associative, inverse, and identity properties of multiplication.

Example 2

Use the multiplication principle to solve each equation

1. $7t = 12$ 2. $\frac{2}{3}x = \frac{1}{4}$ 3. $\frac{3}{w} = \frac{8}{5}$

Solutions

1. $7t = 12$

$$\frac{1}{7} \cdot 7t = \frac{1}{7} \cdot 12 \qquad \text{Multiply both sides by the reciprocal of 7.}$$

$$\left(\frac{1}{7} \cdot 7\right)t = \frac{12}{7} \qquad \text{Simplify the left-hand side using the associative, inverse, and identity properties. Evaluate the rational product on the right-hand side to complete the solution.}$$

$$1 \cdot t = \frac{12}{7}$$

$$t \qquad \frac{12}{7}$$

Let's check our solution by substituting it for t in the original equation.

$$7 \cdot \frac{12}{7} = 12$$

$$\frac{7}{1} \cdot \frac{12}{7} = 12$$

$$\frac{12}{1} = 12 \checkmark \qquad \text{It checks!}$$

2. $\frac{2}{3}x = \frac{1}{4}$

$$\frac{3}{2} \cdot \frac{2}{3}x = \frac{3}{2} \cdot \frac{1}{4}$$
Multiply both sides of the equation by the reciprocal of $\frac{2}{3}$.

$$\left(\frac{3}{2} \cdot \frac{2}{3}\right)x = \frac{3}{8}$$
Simplify the left-hand side using the associative, inverse and identity properties. Evaluate the rational product on the right-hand side to complete the solution.

$$1 \cdot x = \frac{3}{8}$$

$$x = \frac{3}{8}$$

Let's check our work.

$$\frac{2}{3} \cdot \frac{3}{8} = \frac{1}{4}$$

$$\frac{6}{24} = \frac{1}{4}$$

$$\frac{1}{4} = \frac{1}{4} \checkmark$$
It checks!

4. $\frac{3}{w} = \frac{8}{5}$

$$3 \cdot \frac{1}{w} = \frac{8}{5}$$
Rewrite the division as an equivalent multiplication.

$$3 \cdot \frac{1}{w} \cdot w = \frac{8}{5} \cdot w$$
Multiply both sides by the reciprocal of $\frac{1}{w}$

$$3 \cdot \left(\frac{1}{w} \cdot w\right) = \frac{8}{5} \cdot w$$
Simplify the left-hand side using the associative, inverse, and identity properties.

$$3 \cdot 1 = \frac{8}{5} \cdot w$$

$$3 = \frac{8}{5} \cdot w$$

$$\frac{5}{8} \cdot 3 = \frac{5}{8} \cdot \frac{8}{5} \cdot w$$
Multiply both sides by the reciprocal of $\frac{8}{5}$ to complete the solution.

$$\frac{15}{8} = w$$

Now check your work.

$$\frac{3}{\left(\frac{15}{8}\right)} = \frac{8}{5}$$

$$3 \div \frac{15}{8} = \frac{8}{5}$$

$$3 \cdot \frac{8}{15} = \frac{8}{5}$$

$$\frac{3}{1} \cdot \frac{8}{15} = \frac{8}{5}$$

$$\frac{3 \cdot 8}{3 \cdot 5} = \frac{8}{5} \checkmark \qquad \text{It checks!}$$

Practice A

Use either the addition or multiplication principle to solve each equation. Turn the page to check your solutions.

1. $\frac{5}{12} + z = \frac{7}{8}$

2. $8x = \frac{12}{7}$

3. $-\frac{7}{15}q = \frac{11}{6}$

4. $\frac{8}{w} = \frac{5}{3}$

5. $\frac{8}{5} - t = \frac{2}{3}$

6. $\frac{y}{6} = \frac{15}{14}$

B. Two-Step Equations

When the variable is a part of two or more operations, those operations must be undone in the reverse order, like taking off your shoes before you take of your socks. We used this same analogy in Section 2.4. Now, we can use the addition principle to undo additions and the multiplication principle to undo multiplications.

Example 3

Solve each equation.

1. $\frac{2}{5} \cdot q - \frac{1}{3} = \frac{7}{10}$

2. $\frac{3}{8}\left(\frac{1}{3} - y\right) = \frac{5}{4}$

Solutions

1. $\frac{2}{5} \cdot q - \frac{1}{3} = \frac{7}{10}$

$$\frac{2}{5} \cdot q + \left(-\frac{1}{3}\right) = \frac{7}{10}$$ Rewrite the subtraction as an equivalent addition.

$$\frac{2}{5} \cdot q + \left(-\frac{1}{3}\right) + \frac{1}{3} = \frac{7}{10} + \frac{1}{3}$$ Undo the addition first, using the addition principle.

$$\frac{2}{5} \cdot q = \frac{7}{10} + \frac{1}{3}$$

$$\frac{2}{5} \cdot q = \frac{21}{30} + \frac{10}{30}$$

$$\frac{2}{5} \cdot q = \frac{31}{30}$$

$$\frac{5}{2} \cdot \frac{2}{5} \cdot q = \frac{5}{2} \cdot \frac{31}{30}$$ Now undo the multiplication, using the multiplication principle.

$$q = \frac{155}{60}$$ Reduce your answer.

$$q = \frac{31}{12}$$

Check your work.

$$\frac{2}{5} \cdot \frac{31}{12} - \frac{1}{3} = \frac{7}{10}$$

$$\frac{31}{30} - \frac{1}{3} = \frac{7}{10}$$

$$\frac{31}{30} - \frac{10}{30} = \frac{7}{10}$$

$$\frac{21}{30} = \frac{7}{10}$$

$$\frac{7}{10} = \frac{7}{10} \checkmark$$ It checks!

2. $\frac{3}{8}\left(\frac{1}{3} - y\right) = \frac{5}{4}$

$$\frac{3}{8}\left(\frac{1}{3} - y\right) = \frac{5}{4}$$

$$\frac{8}{3} \cdot \frac{3}{8}\left(\frac{1}{3} - y\right) = \frac{8}{3} \cdot \frac{5}{4}$$ Undo the multiplication first because the subtraction is grouped. Use the multiplication principle. Then reduce.

$$\frac{1}{3} - y = \frac{10}{3}$$

$$\frac{1}{3} + (-y) + y = \frac{10}{3} + y$$ Rewrite the subtraction as an equivalent addition, and then use the addition principle. Solve and reduce your answer.

$$\frac{1}{3} = \frac{10}{3} + y$$

$$-\frac{10}{3} + \frac{1}{3} = -\frac{10}{3} + \frac{10}{3} + y$$

$$-\frac{9}{3} = y$$

$$-3 = y$$

The solution is $y = -3$. Let's check it.

$$\frac{3}{8}\left(\frac{1}{3} - (-3)\right) = \frac{5}{4}$$

$$\frac{3}{8}\left(\frac{1}{3} + 3\right) = \frac{5}{4}$$

$$\frac{3}{8}\left(\frac{1}{3} + \frac{9}{3}\right) = \frac{5}{4}$$

$$\frac{3}{8} \cdot \frac{10}{3} = \frac{5}{4}$$

$$\frac{30}{24} = \frac{5}{4}$$

$$\frac{5}{4} = \frac{5}{4} \checkmark$$ It checks!

Practice B

Solve the equations. Turn the page to check your solutions.

7. $8x + 6 = 56$

8. $4d + 2 = \frac{7}{3}$

9. $\frac{5}{3}w - 4 = \frac{8}{3}$

10. $\frac{2}{5}m + \frac{5}{4} = \frac{17}{12}$

11. $\left(\frac{3}{2} - k\right) \cdot \frac{21}{4} = \frac{7}{8}$

12. $\frac{2}{3} = \frac{5}{12} \cdot \left(\frac{n}{10} - \frac{6}{5}\right)$

C. Equations with Decimals

You learned in Section 3.1 that decimals can be written as fractions using a power of 10 in the denominator. When solving an equation with decimals, sometimes it is easier to convert a decimal into fraction notation.

Example 4

Solve the equation: $3.7x - 5.2 = 3.26$. See if you can do it without a calculator.

Solution

$$3.7x - 5.2 = 3.26$$

$$3.7x + (-5.2) + 5.2 = 3.26 + 5.2$$ Rewrite the subtraction as an equivalent addition. Then use the addition principle to undo the addition.

$$3.7x = 8.46$$

$$\frac{37}{10}x = \frac{846}{100}$$ Convert into fraction notation and solve using the multiplication principle. Reduce your answer.

$$\frac{10}{37} \cdot \frac{37}{10}x = \frac{10}{37} \cdot \frac{846}{100}$$

$$x = \frac{8460}{3700}$$

$$x = \frac{423}{185}$$

The solution is $x = \frac{423}{185}$. Enter the left hand side of this equation into your calculator to verify that the solution is correct:

$$3.7 \cdot \frac{423}{185} - 5.2 = 3.26 ✓$$ It checks!

Practice C

Solve the equations. Present answers in fraction notation. Turn the page to check your solutions.

13. $5.4 = 2.4x$

14. $-0.08k = 1.3$

15. $1.2t - 4.6 = 0.8$

16. $2(w - 1.8) = 0.7$

17. $5.3 - 1.9x = 0.74$

18. $(4.5 + r)3.2 = 3.84$

Practice A — Answers

1. $z = \frac{11}{24}$

2. $x = \frac{3}{4}$

3. $q = -\frac{55}{14}$

4. $w = \frac{24}{5}$

5. $t = \frac{14}{15}$

6. $y = \frac{45}{7}$

Exercises 3.8

Solve each equation. When an answer is not a whole number, write it in fraction notation.

1. $w - \frac{1}{3} = \frac{5}{3}$

2. $t + \frac{4}{9} = \frac{7}{9}$

3. $m + \frac{3}{4} = \frac{1}{4}$

4. $n - \frac{5}{3} = \frac{4}{3}$

5. $\frac{5}{3} - p = -\frac{1}{3}$

6. $l + \frac{7}{8} = \frac{1}{8}$

7. $\frac{11}{12} + g = \frac{15}{16}$

8. $\frac{6}{25} - r - \frac{12}{20}$

9. $14 = \frac{7}{5} - z$

10. $5 = -\frac{6}{7} + s$

11. $\frac{4}{5} = \frac{8}{15}w$

12. $9 = -\frac{27}{2}m$

13. $5 = \frac{10}{13}k$

14. $9 = \frac{54}{7}y$

15. $9x = \frac{12}{5}$

16. $-8b = \frac{20}{21}$

17. $-\frac{n}{15} = \frac{7}{12}$

18. $\frac{v}{21} = \frac{13}{14}$

19. $\frac{3}{4} = \frac{y}{8}$

20. $\frac{1}{14} = \frac{k}{21}$

21. $\frac{3}{13} = \frac{3x}{78}$

22. $\frac{4}{15} = -\frac{t}{20}$

23. $\frac{9}{t} = -\frac{12}{5}$

24. $\frac{15}{k} = \frac{25}{16}$

25. $\frac{26}{15} = \frac{13}{r}$

26. $-6.2 = -2.5q$

27. $-3.2y = -0.75$

28. $0.015g = 4.35$

29. $\frac{2}{9}p + \frac{5}{12} = \frac{11}{18}$

30. $\frac{6}{7}r - \frac{5}{14} = \frac{10}{21}$

31. $\frac{3}{2} = \frac{5}{8}x - \frac{7}{4}$

32. $\frac{5}{21} = \frac{t}{6} - \frac{2}{9}$

33. $\frac{6}{5} = \frac{4}{15}r - \frac{3}{10}$

34. $\frac{5}{2} = \frac{3}{4}q - \frac{3}{2}$

35. $\frac{9}{2}v + 2 = \frac{5}{8}$

36. $3s - \frac{4}{9} = \frac{5}{18}$

37. $2m + \frac{4}{3} = \frac{6}{3}$

38. $\frac{4}{5}w + 5 = \frac{10}{3}$

39. $\frac{5}{6}b + \frac{1}{9} = 4$

40. $3d + 2 = \frac{4}{3}$

41. $\frac{5}{4} = \left(v + \frac{3}{2}\right) \cdot \frac{7}{6}$

42. $\frac{5}{6} = \left(w - \frac{3}{2}\right) \cdot 3$

43. $\left(v - \frac{7}{9}\right) \cdot \frac{8}{15} = \frac{4}{25}$

44. $\frac{7}{16}\left(k + \frac{5}{12}\right) = \frac{11}{18}$

45. $\frac{5}{12} = \left(\frac{3}{2} + \frac{k}{3}\right) \cdot \frac{3}{2}$

46. $\frac{9}{8} = \frac{9}{5} \cdot \left(\frac{y}{4} - \frac{9}{2}\right)$

47. $\frac{2}{5}\left(\frac{w}{3} - \frac{7}{15}\right) = \frac{8}{9}$

48. $\left(\frac{1}{2} + \frac{n}{5}\right)\frac{2}{5} = \frac{8}{15}$

49. $5.3t - 7.21 = 0.94$

50. $3(x + 1.2) = 0.9$

51. $5.6 = 3(x + 1.2)$

52. $3.7 = 1.95t - 0.2$

53. $-0.702 = 2.34(1.2 - q)$

54. $6.3 = (w - 2.2) \cdot 6$

55. $(s - 3.07) \cdot 2 = -7.74$

56. $4.7(3.2 - q) = 0.94$

Practice C — Answers

13. $x = \frac{9}{4}$

14. $k = \frac{65}{4}$

15. $t = \frac{9}{2}$

16. $w = \frac{43}{20}$

17. $x = \frac{12}{5}$

18. $r = \frac{-33}{10}$

CHAPTER 4
Expressions

What is a Number?

The English language is full of figures of speech — such as "beating around the bush" — that we use to explain ideas. These are phrases that have meaning beyond their literal, word-for-word interpretation. When we say we're "beating around the bush," we mean avoiding a major, important topic by addressing minor, unimportant topics instead. If someone asks us what we're beating with, we say something like "Oh, that's just an *expression*," a combination of words that work together to create meaning.

In math, an expression is a combination of numbers, variables, and operation symbols that together have meaning. In math, the meaning of an expression is a number. In this chapter, we will distinguish between **arithmetic expressions**, which are expressions that don't contain variables, and **algebraic expressions**, which do contain variables.

4.1 Introduction to Expressions

In this section, you will learn how to evaluate arithmetic and algebraic expressions. You will also learn how to represent and interpret measurements in scientific notation. Finally, we will officially introduce formulas and learn two important ones — the area formula for a rectangle and the area formula for a triangle.

A. Arithmetic Expressions

An arithmetic expression is a combination of numbers and operation symbols. This can be as simple as a single number, such as 23, or it can be a much more complicated combination of numbers and operation symbols, such as $\left(\frac{1}{3} + \frac{29}{21}\right) \cdot \frac{35}{4} + \left(\frac{17-3}{7}\right)^3$.

An arithmetic expression can always be evaluated by performing each of the operations correctly and in the correct order. Once an arithmetic expression has been evaluated, it can then be replaced with its number value. Both the original expression and its evaluated form have the same meaning.

> **Example 1**
>
> Evaluate the arithmetic expressions
>
> 1.   17
> 2. $\frac{3}{4} + \frac{3}{5}$
> 3. $\left(\frac{1}{3} + \frac{29}{21}\right) \cdot \frac{35}{4} + \left(\frac{17-3}{7}\right)^3$
>
> **Solutions**
>
> 1. 17 is an expression that is simply a number.
> 2. $\frac{3}{4} + \frac{3}{5}$
>
> $$\frac{3}{4} + \frac{3}{5} = \frac{15}{20} + \frac{12}{20}$$ The two fractions should be added together.
>
> $$= \frac{27}{20}$$ We know from our experience adding fractions that our evaluation now has the same meaning as our original expression.

3. $\left(\frac{1}{3} + \frac{29}{21}\right) \cdot \frac{35}{4} + \left(\frac{17-3}{7}\right)^3$

$\left(\frac{1}{3} + \frac{29}{21}\right) \cdot \frac{35}{4} + \left(\frac{17-3}{7}\right)^3 = \frac{12}{7} \cdot \frac{35}{4} + 2^3$ In this expression, we have some operations to complete before others in our evaluation. Parentheses first.

$\frac{12}{7} \cdot \frac{35}{4} + 2^3 = \frac{12}{7} \cdot \frac{35}{4} + 8$ Next is the exponent.

$\frac{12}{7} \cdot \frac{35}{4} + 8 = 15 + 8$ Now we can multiply our fractions and simplify the result.

$15 + 8 = 23$ One last bit of addition gives us our answer! The number 23 and our original expression have the same meaning, but they certainly don't look very similar.

Practice A

Evaluate the following arithmetic expressions. Turn the page to check your solutions.

1. $\frac{17}{13}$

2. $2(3)^2$

3. $\frac{3}{5} + \frac{1}{3} \cdot \frac{6}{5}$

4. $-2^4 + (-3)^2 - (-2)^6$

5. $\dfrac{-2^2 + 3 + (-3) + 4}{(4-5)(-2-7)}$

6. $\dfrac{15 - 7(2+4)}{3^3}$

B. Scientific Notation

In Example 1, we saw that the complicated arithmetic expression $\left(\frac{1}{3} + \frac{29}{21}\right) \cdot \frac{35}{4} + \left(\frac{17-3}{7}\right)^3$ and the much simpler arithmetic expression 23 are actually equivalent. When we evaluate the first, we end up with the second. It's hard to imagine a scenario where we would *prefer* to use the expression $\left(\frac{1}{3} + \frac{29}{21}\right) \cdot \frac{35}{4} + \left(\frac{17-3}{7}\right)^3$ instead of the number 23, but believe it or not, there are times when it makes sense to use a more complex expression in place of a number.

For example, suppose you drive a dump truck full of gravel onto a scale that shows your truck weighs 78,000 pounds. Simply writing out the measurement as a number fails to communicate the accuracy of the measurement. Is it accurate to the nearest pound? To the nearest 10 pounds? To the nearest 100 pounds? To the nearest 1,000 pounds? Any of these are possibilities when a scale measures weights that massive.

If the measurement is accurate to the nearest 1,000 pounds, then we only know that the actual weight is at least 77,500 pounds but less than 78,500 pounds. The precise weight could actually be any number in that range. None of the zeros in 78,000 are significant because the digits in those positions are entirely unknown. They are merely placeholders.

If the measurement is accurate to the nearest 100 pounds, then we know that the actual weight is at least 77,950 but less than 78,050. It could be any number in that range, but this is now a smaller range of uncertainty. The zero in the hundreds place of 78,000 is significant because we now know that the hundreds place digit must be either 0 or 9, but the zeros in the ones and tens places are placeholders.

Using scientific notation to write measurements like this allows us to communicate which zeros are place holders and which are significant. This shows others the level of precision in the measurement. Here is how we would write this measurement in scientific notation in each scenario:

- 7.8×10^4: The only significant digits are the 7 and the 8. This measurement has two *significant figures*, which is often shortened to "sig figs," of precision.

- 7.80×10^4: This measurement has three sig figs. Because the zero that is two places to the right of the decimal point is included, we know that it is significant.

In both cases, the expression evaluates to 78,000. Using scientific notation allows us to specify whether the zero in the hundreds place is significant or merely a placeholder.

Writing Numbers in Scientific Notation

Here are the steps for writing a number in scientific notation:

1. Write out all of the significant figures, but don't write place-holder zeros.

2. If there is more than one significant figure, place a decimal point between the left-most digit and the digit to its right.

3. Multiply by the appropriate power of 10. The exponent should be the number of places the decimal point must move to reach it's new position. (Note: If the original number has no decimal point then the exponent is one less than the number of digits in the original number.)

Example 2

Dinesh competes in long distance bicycle tournaments, and he measures the length of his most recent bicycle ride to four significant figures as 128,000 meters. Write this measurement in scientific notation.

Solution
Following the steps above:

1. We are told that there are four significant figures. This gives us 1280.

2. We place the decimal point between the 1 and the 2: 1.280

3. The orginal number, 128,000, has no decimal point. Since it has six digits, the exponent should be 5: one less than 6. We end up with 1.280×10^5.

Example 3

Liesel uses an electronic balance to measure the mass of a sample of the mineral cobalt during a chemistry experiment. The readout is in scientific notation: 2.300×10^2 grams.

1. How many significant figures does this measurement have?

2. Write this measurement in standard notation.

3. What is the range of possibilities for the actual mass of the sample?

Solution

1. There are four significant figures. The zeros would have been omitted if they were place-holders, so they must be significant.

2. Multiplying by 10^2 means "multiply by 10 two times." Each time we multiply by 10, the decimal point shifts one place to the right, so $2.300 \times 10^2 = 230.0$. We may write this as a whole number, 230, but then we lose sight of the fact that the measurement is accurate to the nearest tenth of a gram.

3. The actual mass is at least 229.5 grams but less than 230.5 grams.

Practice B

Write each number in scientific notation with 5 significant figures.

7. 247,000,000   8. 247,139,812   9. 2471.398

Write each number in standard notation and state the number of significant figures recorded.

10. 1.3400×10^{11} 11. 1.34×10^{11} 12. 8.53129×10^4

When you are finished, turn the page to check your solutions.

Practice A — Answers

1. $\frac{17}{13}$ 4. -71
2. 18   5. 0
3. 1   6. -1

C. Algebraic Expressions

When a mathematical expression contains a variable, then it is no longer an arithmetic expression. It's an algebraic expression, instead. We may use a single algebraic expression to represent an entire class of calculations. These are called **formulas**. When we already know that variables in certain expressions will always tell us something, such as the area of a rectangle, solving problems like this can be much easier!

Remember that a rectangle that measures l units long and w units wide has an area of $l \cdot w$ square units. The algebraic expression $l \cdot w$ can be used to represent the area of *any* rectangle. To find the area of a particular rectangle, we replace the variables l and w with the measurement of the length and width for that rectangle. After replacing the variables with the appropriate number values, the *algebraic* expression becomes an *arithmetic* expression, which we can then proceed to evaluate. This process is called **evaluating an algebraic expression**.

Example 4

Find the area of a rectangle that measures 6.2 meters long and 2.5 meters wide.

Solution

We evaluate the algebraic expression $l \cdot w$ for $l = 6.2$ and $w = 2.5$ to find the area of the rectangle: The area is $6.2 \cdot 2.5 = 15.5$ square meters.

Let's look at another polygon. A triangle is a polygon with three sides. We need two measurements to calculate the area of a triangle, the *base* and *height*. The base can be any of the three side lengths. The height is the shortest distance from the line that goes through the base to the point where the other two sides meet.

The formula is different for finding the area of a triangle. The area of a triangle with a base measurement of b units and a height measurement of h units is $\frac{1}{2} \cdot b \cdot h$.

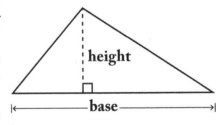

Figure 1

Example 5

Find the area of a triangle that has base $\frac{10}{3}$ inches and height $\frac{9}{5}$ inches.

Solution

We evaluate the algebraic expression $\frac{1}{2} \cdot b \cdot h$ for $b = \frac{10}{3}$ and $h = \frac{9}{5}$ to find the area of the triangle. The area is $\frac{1}{2} \cdot \frac{10}{3} \cdot \frac{9}{5} = \frac{90}{30} = 3$ square inches.

Example 6

The expression $\frac{9000n \cdot 1.005^{12n}}{1.005^{12n} - 1}$ represents the total repayment amount for a \$150,000 mortgage at 6% APR (usually called the "interest rate") if it is paid off in equal monthly installments over n years.

1. Evaluate this expression for $n = 30$. Round your answer to the nearest dollar.

2. Evaluate this expression for $n = 15$. Round your answer to the nearest dollar.

3. How much more will a 30-year loan cost than a 15-year loan?

Solutions

We can use a calculator to evaluate these expressions:

1. $\frac{9000 \cdot 30 \cdot 1.005^{12 \cdot 30}}{1.005^{12 \cdot 30} - 1} \approx 323{,}757$

2. $\frac{9000 \cdot 15 \cdot 1.005^{12 \cdot 15}}{1.005^{12 \cdot 15} - 1} \approx 227{,}841$

3. The 30-year loan will cost \$323,757 – \$227,841 = \$95,916 more than a 15-year loan.

Practice C

Find the area of the following polygons. Turn the page to check your solutions.

13. Find the area of a rectangle with length $\frac{9}{2}$ inches and width $\frac{45}{16}$ inches.

14. Find the area of a triangle with base 9 meters and height $\frac{8}{3}$ meters.

15. Find the area of the rectangle:

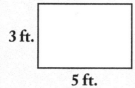

3 ft.

5 ft.

17. Find the area of the triangle:

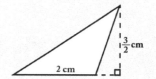

$1\frac{3}{2}$ cm

2 cm

16. Find the area of the triangle:

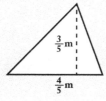

$\frac{3}{5}$ m

$\frac{4}{5}$ m

18. Find the area of the triangle:

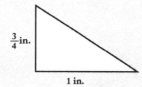

$\frac{3}{4}$ in.

1 in.

D. Areas of Composite Figures

With the information about areas of rectangles and triangles, you can now find the areas of composite figures. A **composite figure** is the result of combining two or more **simple figures** — for now, just rectangles and triangles — either by combining them together or removing one from the other. The composite figures in both examples below use the same two simple figures: a rectangle with area 8 square centimeters and a triangle with area 24 square centimeters.

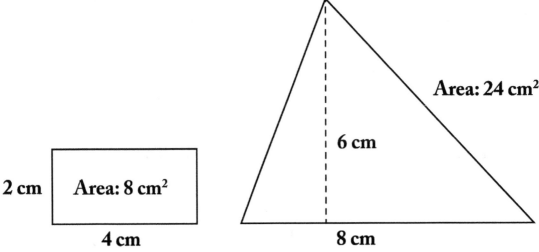

Figure 2. Note: Figures are not to scale.

When we glue two simple figures together to create a composite figure, then the area of the composite figure is the sum of the areas of the simple figures.

Example 7

Find the area of the composite figure.

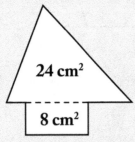

Solution

In this polygon, we can combine the two areas to evaluate the total area of the polygon.

Total area: 24 cm² + 8 cm² = 32 cm²

When we remove one simple figure from another simple figure to create a composite figure, then the area of the composite figure is the difference of the areas of the simple figures.

Example 8

Find the area of the composite figure.

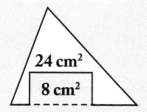

Solution

In this polygon, we can subtract the missing area of the rectangle from the area of the triangle.
Total Area: 24 cm^2 – 8 cm^2 = 16 cm^2

Formulas like these create easily recognizable expressions that can be used to solve simple problems like area. In the exercises for this section, you'll find more complex formulas for finding the area of more shapes.

Practice B — Answers

7. 2.4700×10^8

8. 2.4714×10^8

9. 2.4714×10^3

10. 5 significant figures: 134,000,000,000

11. 3 significant figures: 134,000,000,000

12. 6 significant figures: 85,312.9

Practice D

Find the area of the following figures. Turn the page to check your solutions.

19. Find the area of the composite figure:

20. Find the area of the composite figure:

21. Find the area of the composite figure:

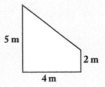

22. Find the area of the composite figure:

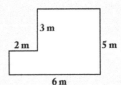

23. Find the area of a three foot wide concrete walkway around the outside edge of a rectangular rose garden as shown, if the rose garden is 21 feet long by 15 feet wide.

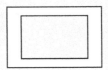

24. Find the area of the diamond, imagining it as two triangles put together.

Exercises 4.1

For the following exercises, classify each expression as either *arithmetic* or *algebraic*.

1. $17t - 12w$

2. 5.32×10^7

3. $\dfrac{2 - \frac{7}{3}}{\frac{11}{4} + \frac{1}{3}}$

4. $\dfrac{9v + 5}{v - 3}$

5. x

6. mc^2

7. $3x$

8. 2

For the following exercises, write each measurement in standard notation and state the number of significant figures that are represented.

9. 3.43×10^2 meters per second (speed of sound in dry air)

10. 1.497×10^3 meters per second (speed of sound in water at 25 degrees Celsius)

11. 2.389×10^5 miles (approximate distance to the moon)

12. 9.296×10^7 miles (approximate distance to the sun)

13. 4.370×10^6 kilometers (approximate circumference of the sun as in Figure 3)

Figure 3.

14. 4.02×10^{13} kilometers (approximate distance to Proxima Centauri, the next closest star after the sun)

15. 3.00×10^8 meters per second (speed of light)

16. 7.0×10^8 meters (average radius of the sun as in Figure 3)

For the following exercises, write each measurement in scientific notation representing the indicated number of significant figures.

17. 3400 kilometers (average radius of Mars: use 2 significant figures)

18. 58,000 kilometers (average radius of Saturn: use 2 significant figures)

19. 1,988,500,000,000,000,000,000,000,000,000 kilograms (approximate mass of the sun: use 4 significant figures)

20. 382,800,000,000,000,000,000,000,000 watts (approximate luminosity of the sun: use 4 significant figures)

21. 1,083,210,000,000 cubic kilometers (approximate volume of the earth: use 6 significant figures)

22. 40,000 kilometers (approximate circumference of Earth: use 2 significant figures)

23. 86,400 seconds (number of seconds in a day: use 5 significant figures)

24. 140,000,000 meters (diameter of Jupiter: use 3 significant figures)

For the following exercises, evaluate the area of each figure.

25. Rectangle: length is 7 feet, and width is 3 feet.

26. Triangle: base is 18 meters, and height is 5 meters.

27. Rectangle: length is $\frac{26}{5}$ inches, and width is $\frac{8}{13}$ inches.

28. Triangle: base is $\frac{15}{4}$ cm, and height is $\frac{16}{9}$ cm.

29. Triangle: base is $\frac{7}{2}$ meters, and height is 12 meters.

30. Rectangle: length is $\frac{12}{5}$ yards, and width is 10 yards.

31. Triangle: base is $\frac{12}{7}$ miles, and height is $\frac{14}{3}$ miles.

32. Rectangle: length is $\frac{24}{11}$ inches, and width is $\frac{9}{16}$ inches.

For the following exercises, evaluate each expression.

33. Area of a trapezoid with height h and base lengths a and b: $\frac{h(a+b)}{2}$ for $h = 7, a = 11$, and $b = 17$

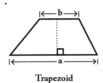

Trapezoid

34. Surface area of a rectangular box measuring l units long, w units wide and h units high: 2 $(lw + wh + lh)$ for $l = 25, w = 8$, and $h = 4$.

35. Balance on a $150,000 fixed-rate mortgage at 6% APR after n years with monthly payments of $P: $150,000 \cdot 1.005^{12n} - 200P(1.005^{12n} - 1)$ for $n = 10$ and $P = 900$.

36. Balance on a $150,000 fixed-rate mortgage at 6% APR after n years with monthly payments of $P: $150,000 \cdot 1.005^{12n} - 200P(1.005^{12n} - 1)$ for $n = 20$ and $P = 900$.

37. Percent return on an investment with future value F and present value P: $\frac{100(F-P)}{P}$ for $F = 2400$ and $P = 1000$.

38. Percent return on an investment with future value F and present value P: $\frac{100(F-P)}{P}$ for $F = 2400$ and $P = 800$.

19. 12 ft^2

20. 6 ft^2

21. 14 m^2

22. 24 m^2

23. 252 ft^2

24. $\frac{21}{2}$ m^2

39. Monthly payment for a $150,000 fixed-rate mortgage at 6% APR for n years: $\frac{750 \cdot 1.005^{12n}}{1.005^{12n} - 1}$ for $n = 30$.

40. Monthly payment for a $150,000 fixed-rate mortgage at 6% APR for n years: $\frac{750 \cdot 1.005^{12n}}{1.005^{12n} - 1}$ for $n = 15$.

41. Remaining distance of a marathon after running t minutes at a pace of P minutes per mile: $26.2 - \frac{t}{P}$ for $t = 140$ and $P = 7$ minutes per mile.

42. Remaining distance of a marathon after running t minutes at a pace of P minutes per mile: $26.2 - \frac{t}{P}$ for $t = 140$ and $P = 8$ minutes per mile.

Find the areas of the composite figures.

43.

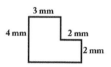

48.

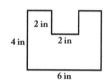

44.

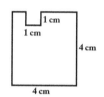

49.

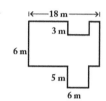

45.

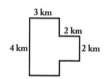

50.

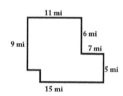

46.

51.

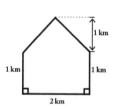

47.

52.

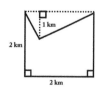

53.

54.

55.

56.

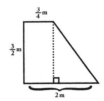

57.

58.

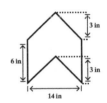

4.2 Combining Like Terms

The **perimeter** of a closed geometric figure is the total distance around the outside edge of the figure. If the figure is a **polygon**, which is a closed geometric figure with straight sides, then the perimeter is the sum of the lengths of the sides. A rectangle is a polygon with four sides that meet at right angles. Two have length l, and the other two have length w.

We will explore three different expressions for the perimeter of a rectangle. Each expression comes from a slightly different way of thinking about the calculation, and each is correct.

1. Start at the upper left-hand corner and move completely around the rectangle in the clockwise direction, adding up the side measurements as you go: $l + w + l + w$.

2. Start at the upper left-hand corner and move *half* way around the rectangle in the clockwise direction, adding up the side lengths as you go. This will be *half* the perimeter, so multiply the result by two to get the full perimeter: $2(l + w)$.

3. Start with the first expression $l + w + l + w$. Because of addition's commutative and associative properties, we can add these four side lengths in any order we please. We first reverse the order of the middle two, $l + l + w + w$, and then use multiplication to represent the repeated additions: $2l + 2w$.

A. The Distributive Property

The three expressions above are all **equivalent** to each other. That means that whenever you evaluate these three expressions using the same substituted numbers, the resulting values are guaranteed to be equal to each other. For example, if $l = 9$ and $w = 4$, then:

- $l + w + l + w = 9 + 4 + 9 + 4 = 13 + 9 + 4 = 22 + 4 = 26$
- $2(l + w) = 2(9 + 4) = 2 \cdot 13 = 26$
- $2l + 2w = 2 \cdot 9 + 2 \cdot 4 = 18 + 8 = 26$

If we use any other substitution, we will have similar results: the three expressions will result in matching values. That is because they are equivalent expressions.

In fact, there is an important property of algebra that ties the second and third expressions together, called the **distributive property of multiplication over addition**.

Distributive Property of Multiplication Over Addition

Given any three numbers a, b, and c, the expressions $a(b + c)$ and $ab + ac$ are equivalent.

There are two ways we can use the distributive property. Replacing an expression of the form $a(b + c)$ with the equivalent expression of the form $ab + ac$ is called *distributing*, or *multiplying*. Replacing an expression of the form $ab + ac$ with the equivalent expression of the form $a(b + c)$ is called *factoring*.

- $a(b + c) \rightarrow ab + ac$: distributing or multiplying
- $ab + ac \rightarrow a(b + c)$: factoring

Example 1

Distribute each expression.

1. $2(l + w)$
2. $-5(3x - 7)$
3. $(3 + 2n)4n$
4. $2 - (8 + r)$

Solutions

1. $2(l + w) = 2l + 2w$
2. $-5(3x - 7) = -5 \cdot 3x - (-5) \cdot 7 = -15x + 35$
3. $(3 + 2n)4n = 3 \cdot 4n + 2n \cdot 4n = 12n + 2 \cdot 4 \cdot n \cdot n = 12n + 8n^2$
4. $2 - (8 + r) = 2 + (-1)(8 + r) = 2 + (-1) \cdot 8 + (-1) \cdot r = 2 + (-8) + (-r) = -6 - r$

Notice that in the last three examples, we were able to evaluate some of the operations in the expression after we distributed, so we ended up with an expression that is equivalent to the original expression but less complicated. This process is called **simplifying** an expression.

Distributing may allow us to simplify an expression that can't otherwise be simplified. Based on order of operations, for example, we can't evaluate any other operations in the expression $-5(3x - 7)$ until we evaluate the multiplication $3x$. That means that without the distributive property, we would be stuck! We can't simplify this expression because we can't evaluate the operation that must come first. However, after we distribute the $-5(3x - 7)$, we can simplify to get $-15x + 35$.

Example 2

Factor each expression

1. $2l + 2w$
2. $5k - 12k$
3. $9w^2 + 27w$

Solutions

1. $2l + 2w = 2(l + w)$

2. $5k - 12k = (5 - 12)k = -7k$

3. $9w^2 + 27w = 9w \cdot w + 9w \cdot 3 = 9w(w + 3)$

In the second problem of Example 2, factoring allows us to simplify an expression that could not otherwise be simplified. Based on order of operations, we can't evaluate the subtraction in the expression $5k - 12k$ until both multiplications (by k) have been evaluated. Again, without the distributive property, we are stuck. But after factoring, we can simplify the equivalent expression $(5 - 12)k$ to obtain $-7k$.

Practice A

Either distribute or factor, and then simplify. Turn the page to check your solutions.

1. $2(x + y)$
2. $2x + kx$
3. $(2s + 5t)8$

4. $9h + 11h$
5. $-(x + y)$
6. $-(m - n)$

B. Combining Like Terms

The **terms** of an expression are the parts of the expression that are separated by ungrouped additions after all ungrouped subtractions have been rewritten as equivalent additions. For example, the terms of the expression $7ab + \frac{3a}{a-b} - 9a + 3ab + a$ are are:

- $7ab = 7 \cdot ab$
- $3a\_a - b = 3 \cdot a\_a - b$
- $-9a = -9 \cdot a$
- $3ab = 3 \cdot ab$
- $a = 1 \cdot a$

Each term in the expression has a **constant factor**. In these examples, the constant factor is the number between the equal sign and the multiplication operator. If the constant factor is 1 or −1, then it can be left out when the expression is written. This is the case in the last term of the above expression.

Also, each term *might have* a **variable factor**. In these examples, the variable factor is the part that is to the right of the multiplication operator. If the term has no variable factor, then it is either a simple number, or it can be simplified to a simple number. All of the operations in the term can be evaluated since there is no variable.

Did you notice that the first and fourth terms have the *same* variable factor, ab? The third and fifth terms have the same variable factor, a. In an expression, terms that have the same variable factor are called **like terms**, and they can be combined to simplify the expression.

To combine like terms, follow these steps:

1. Rewrite any ungrouped subtractions as additions.

2. Change the order of the terms so that like terms are next to each other. Step two cannot be completed until step one is completed because subtraction is neither commutative nor associative.

3. Combine like terms by factoring out the variable factor and evaluating the addition.

Example 3

Simplify the expression by combining like terms: $7ab + \frac{3a}{a-b} - 9a + 3ab + a$

Solution

Step 1	$7ab + \frac{3a}{a-b} + (-9a) + 3ab + a$	Rewrite any ungrouped subtractions as additions.
Step 2	$7ab + 3ab + (-9)a + 1 \cdot a + \frac{3a}{a-b}$	Change the order of the terms so that like terms are next to each other.
Step 3	$(7+3)ab + (-9+1) \cdot a + \frac{3a}{a-b}$	Combine like terms by factoring out the variable factor.
Step 4	$10ab + (-8)a + \frac{3a}{a-b}$	Evaluate the addition.

Example 4

Simplify the following expressions by combining like terms.

1. $3pq - 4p + 2q + 8pq$

2. $5x^2 - x + 7 + x^2 + 4x$

Solutions

1. $3pq - 4p + 2q + 8pq$

$3pq - 4p + 2q + 8pq = 3pq + (-4p) + 2q + 8pq$
Rewrite any ungrouped subtractions as additions.

$$= 3pq + 8pq + (-4p) + 2q$$

Change the order of the terms so that like terms are next to each other.

$$= 11pq + (-4p) + 2q$$

Combine like terms by factoring out the variable factor.

2. $5x^2 - x + 7 + x^2 + 4x$

$$5x^2 - x + 7 + x^2 + 4x = 5x^2 + (-1x) + 7 + 1x^2 + 4x$$

Rewrite any ungrouped subtractions as additions.

$$= 5x^2 + 1x^2 + (-1x) + 4x + 7$$

Combine like terms by factoring out the variable factor.

$$= 6x^2 + 3x + 7$$

Change the order of the terms so that like terms are next to each other.

In the second expression above, there are terms with variable parts (x) that have different powers of the same variable. These are *not* like terms! We end up with $6x^2 + 3x + 7$, and this expression can't be simplified any further because the first and second terms have *different* variable factors.

Practice B

Simplify each expression by combining like terms. Then turn the page to check your solutions.

7. $2xy + 3xy$

8. $11xy^2z + 8xy^2z$

9. $2a + 3b + 4a - b$

10. $8t - t^3 + 7t^2 - 5t + 3t^2$

11. $7m - mn + 3n - 5mn$

12. $-9hp^2 + 4hp + 6h^2p - 5hp^2 + hp$

C. Factoring Out the GCF

In a previous example, the instructions were to factor the expression $9w^2 + 27w$. Our solution looked like this:

$$9w^2 + 27w = 9w \cdot w + 9 \cdot w3 = 9w(w + 3)$$

Practice A — Answers

1. $2x + 2y$

2. $(2 + k)x$

3. $16s + 40t$

4. $20h$

5. $-x - y$

6. $-m + n$

Notice that both terms have a factor of $9w$. In the first term, the other factor is w, and in the second term, the other factor is 3. Because w and 3 have no common divisors, we say that the greatest common factor of $9\,w^2$ and $27w$ is $9w$. In this example, factoring did not allow us to simplify the expression because $9\,w^2$ and $27w$ are not like terms.

Example 5

Factor out the GCF in the following expressions.
1. $36\,d^3 - 24\,d^2$
2. $45a\,b^2 - 30\,a^2 b + 75ab$

Solutions

1. $36\,d^3 - 24\,d^2$

$$36\,d^3 - 24\,d^2 = 12\,d^2 \cdot 3d - 12\,d^2 \cdot 2$$
$$= 12\,d^2(3d - 2)$$

2. $45a\,b^2 - 30\,a^2 b + 75ab$

$$45a\,b^2 - 30\,a^2 b + 75ab = 15ab \cdot 3b - 15ab \cdot 2a + 15ab \cdot 5$$
$$= 15ab(3b - 2a + 5)$$

There are situations in which it will be important to write an equation in factored form, even if doing so keeps you from simplifying the expression further. One such situation involves writing an expression for the area of a trapezoid. A **trapezoid** is a four-sided polygon in which two of the sides are *parallel to one another*. That means the sides are on lines that would never intersect even if they continued infinitely in both directions. The parallel sides are called the **bases** of the trapezoid. The distance between the lines that the bases are on is called the **height** of the trapezoid.

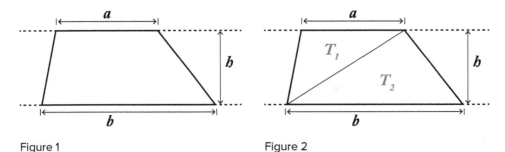

Figure 1   Figure 2

In Figure 1, the lengths of the bases are labeled a and b, and the height is labeled h. The parallel lines that the base sides are on are shown as dotted lines for the sake of illustration. The height is not usually a side length in the trapezoid.

We calculate the area of the trapezoid by dividing it into two triangles, T_1 and T_2, along a diagonal. We can see in Figure 2 that the trapezoid is actually the compound figure that results from combining triangles T_1 and T_2 together. The area of the trapezoid is the sum of the area of the triangles T_1 and T_2. Both triangles have height h. The base measurement for T_1 is a, and the base measurement for T_2 is b, so their areas are $\frac{1}{2}ah$ and $\frac{1}{2}bh$. Adding these areas together give us this expression for the area of the trapezoid:

$$\frac{1}{2}ah + \frac{1}{2}bh$$

When we write an expression for the area of a trapezoid, we usually factor out the GCF, $\frac{1}{2}h$. The resulting expression, $\frac{1}{2}h(a + b)$, has only three operations, so it's easier to evaluate than the equivalent expression $\frac{1}{2}ah + \frac{1}{2}bh$, which has five operations.

Example 6

Find the area of a trapezoid with height $h = \frac{8}{3}$ inches, and base measurements $a = \frac{15}{4}$ inches and $b = \frac{12}{5}$ inches.

Solution

Evaluate the expression $\frac{1}{2}h(a + b)$ for $h = \frac{8}{3}$, $a = \frac{15}{4}$, and $b = \frac{12}{5}$.

$$\frac{1}{2} \cdot \frac{8}{3}\left(\frac{15}{4} + \frac{12}{5}\right) = \frac{1}{2} \cdot \frac{8}{3}\left(\frac{75}{20} + \frac{48}{20}\right)$$

$$= \frac{1}{2} \cdot \frac{8}{3} \cdot \frac{123}{20}$$

$$= \frac{41}{5}$$

The area of the trapezoid is $\frac{41}{5}$ square inches.

Practice C

In each expression, factor out the GCF. Turn the page to check your answers.

13. $6a + 3b$

14. $12m^2 - 8n$

15. $7ab + 11b^2$

16. $12fg + 20gh$

17. $18x + 24x^2$

18. $32tk^2 + 40t^2k - 16tk$

Exercises 4.2

For the following exercises, distribute and then simplify if possible.

1. $4(x + 1)$

2. $3(x - 2)$

3. $4(7x - 3)$

4. $3(5z + 8)$

5. $-(2x + 4)$

6. $-(-3w - 5)$

7. $3 - (5 - t)$

8. $8 - (h + 2)$

9. $g(5 + 2g)$

10. $(6 - 4y)y$

For the following exercises, simplify by combining like terms.

11. $7x + 3x$

12. $13w + 24w$

13. $7uv + 9u - 5v + 3uv - 4u$

14. $9ab^2 + 3ab - 4ab^2 + 6ab$

15. $3n^2 - 8n + 2n - n^2$

16. $k - 5k^2 + 4k - 2k^3$

17. $6h^3 - 2h + h^3 - 3h^2$

18. $11p^2 - 5 + 7p + p^2 - 8$

19. $2u^3v^2 + 5u^2v^3 - 5u^2v^2 + 7u^2v - 3u^2v$

20. $4a^3b^3 + 5a^3b^2 - 4a^3b^3 + a^3b$

For the following exercises, distribute and then combine like terms.

21. $3(x - 1) - 2$

22. $4(w - 2) - 2w + 8$

23. $5(2x - 4) - 3(6 + 3x)$

24. $8(a - 3b) + 5(2a + b)$

25. $2n(3 - 4n) + 6(4 + 2n)$

26. $7(2 - 5p) - 3(2p - 6)$

27. $(2y + 1)y - (3y + 2)3$

28. $3b(a + 2b) - 4a(3a - b)$

For the following exercises, factor out the GCF.

29. $6x - 4$

30. $9a + 12$

31. $4w - 4$

32. $10r - 5$

33. $4x^2 - 5x$

34. $9w^2 - 2w$

35. $11n^3 + 4n^2$

36. $14m^5 - 9m^3$

37. $8x^3 - 12x^2 + 20x$

38. $18m - 54m^2 + 45m^3$

39. $15d^2r + 35dr^2 - 20dr$

40. $21p^2v - 18pv^2 - 12pv$

Think about these problems and then answer them carefully.

41. Here's a new way to think about adding fractions with a common denominator. Earlier in the term we learned that when we are operating in the set of rational numbers, any division can be rewritten as an equivalent multiplication: $\frac{a}{c} = a \cdot \frac{1}{c}$. Consider the sum of fractions $\frac{a}{c} + \frac{b}{c}$ and then do the following:

   a. Rewrite the fractions as equivalent multiplications.

   b. Factor out the GCF

   c. Rewrite the result as a fraction. What do you end up with?

42. Use the expression for the area of a trapezoid to find the area of a trapezoid with base measurements $a = 5x, b = 7x$ and height $h = 2x$. Simplify your answer.

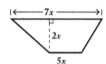

44. Find the area of the trapezoid. Simplify your answer. Give your answer to the hundredth of a centimeter.

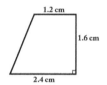

43. Find the area of the trapezoid. Simplify your answer.

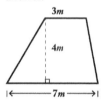

45. A parallelogram can be thought of as a trapezoid in which the base measurements are equal. Simplify the expression for the area of a trapezoid with base measurements b and b and height h to obtain an expression for the area of a parallelogram.

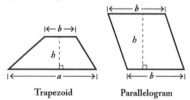

Trapezoid   Parallelogram

46. Find the area of the parallelograms.

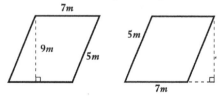

47. The figure below that is outlined in black can be divided into a trapezoid and a rectangle that is missing a triangular piece. Use the measurements that are given and the expressions for the areas of a trapezoid, rectangle, and triangle to write an expression for the area of this figure. Simplify the expression.

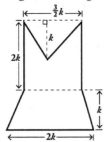

48. Find the area of the composite figure.

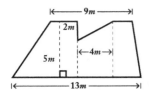

Write an expression for the area of each figure. Use the indicated measurement variables.

49.

52.

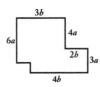

50.

53.

51.

54.

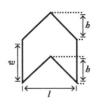

4.3 Simplifying Expressions with Algebraic Fractions

An **algebraic fraction** is a fraction in which either the numerator or denominator is an algebraic expression.

The key to understanding algebraic fractions is to think of each variable as if it were a positive integer. We used this mental trick in the last section to find the greatest common factor of $9\,w^2$ and $27w$. We treated the variable w as if it were a positive integer, even though it could represent any rational number. That's why we included w as a common divisor of $9\,w^2$ and $27w$. Using this logic, GCF$(9w^2, 27w) = 9w$ and lcm$(9\,w^2, 27w) = 27w^2$. We use GCF to reduce fractions, and we use LCM to add fractions with different denominators. Both are important techniques for simplifying expressions with algebraic fractions.

A. Reducing Algebraic Fractions and Their Products

To reduce the algebraic fraction $\frac{9\,w^2}{27w}$, we factor out the GCF of the numerator and denominator then cancel the GCF. This is the same process that we used to reduce fractions in Chapter 3:

$$\frac{9\,w^2}{27w} = \frac{w \cdot 9w}{3 \cdot 9w} = \frac{w}{3}$$

We call this **canceling a common factor** of $9w$. It may be helpful to actually cross out the common factors in your work, like this:

$$\frac{9\,w^2}{27w} = \frac{w \cdot \cancel{9w}}{3 \cdot \cancel{9w}} = \frac{w}{3}$$

In Chapter 3, we saw that even if two fractions are in lowest terms, their product may not be in lowest terms. This is because the numerator of one fraction might have a common factor with the denominator of the other. For example, even though $\frac{3}{4}$ and $\frac{8}{9}$ are both in lowest terms, their production $\frac{3}{4} \cdot \frac{8}{9} = \frac{24}{36}$ is not. We can reduce the product *after* we multiply:

$$\frac{3}{4} \cdot \frac{8}{9} = \frac{24}{36} = \frac{2 \cdot \cancel{12}}{3 \cdot \cancel{12}} = \frac{2}{3}$$

Or we can reduce the product *before* we multiply:

$$\frac{3}{4} \cdot \frac{8}{9} = \frac{3}{4} \cdot \frac{4 \cdot 2}{3 \cdot 3} = \frac{2}{3}$$

Reducing a product of fractions before we multiply is called **cross-canceling** because we are canceling common factors of the numerator and denominator *across* the multiplication symbol.

Example 1

Reduce the following products by cross-canceling.

1. $\frac{2t}{9} \cdot \frac{6}{5t}$

2. $\frac{5p^2}{6n} \cdot \frac{3\,n^2}{10p}$

Solutions

1. $\dfrac{2t}{9} \cdot \dfrac{6}{5t}$

$$\dfrac{2t}{9} \cdot \dfrac{6}{5t} = \dfrac{2 \cdot \cancel{t}}{3 \cdot 3} \cdot \dfrac{2 \cdot 3}{\cancel{t} \cdot 5}$$

By factoring each algebraic fraction that we will multiply, we can cross-cancel the 3 and the t.

$$= \dfrac{4}{15}$$

The result has no variables at all!

2. $\dfrac{5p^2}{6n} \cdot \dfrac{3n^2}{10p}$

$$\dfrac{5p^2}{6n} \cdot \dfrac{3n^2}{10p} = \dfrac{p \cdot \cancel{5p}}{2 \cdot \cancel{3n}} \cdot \dfrac{\cancel{3n} \cdot n}{\cancel{5p} \cdot 2}$$

The same cross-cancelling works in this example.

$$= \dfrac{pn}{4}$$

We're left with variables, but the reduced fraction is much simpler.

⊳ With more complicated algebraic fractions, we may need to factor out the GCF in the numerator and denominator before we reduce the faction.

Example 2

Reduce the following algebraic fractions by first factoring out the GCF from both the numerator and the denominator.

1. $\dfrac{15n^2 - 45n}{10n^2 + 20n}$

2. $\dfrac{3z^2 + 5z}{6z + 10}$

Solutions

1. $\dfrac{15n^2 - 45n}{10n^2 + 20n}$

$$\dfrac{15n^2 - 45n}{10n^2 + 20n} = \dfrac{15n(n - 3)}{10n(n + 2)}$$

$15n$ is a variable factor of both rational numbers in the numerator, and $10n$ is a variable factor of both rational numbers in the denominator. We can factor those out.

$$= \dfrac{\cancel{5n} \cdot 3(n - 3)}{\cancel{5n} \cdot 2(n + 2)}$$

We can factor both of those further to reach a GCF of $5n$ in the numerator and denominator, which can be canceled.

$$= \dfrac{3(n - 3)}{2(n + 2)}$$

We are left with a much simpler fraction.

2. $\dfrac{3z^2 + 5z}{6z + 10}$

$\dfrac{3z^2 + 5z}{6z + 10} = \dfrac{z\cancel{(3z+5)}}{2\cancel{(3z+5)}}$

z is a variable factor of both rational numbers in the numerator, and 2 is a constant factor of both rational numbers in the denominator. Since the remaining parenthetical group is identical in both, they can be canceled.

$= \dfrac{z}{2}$

Leaving us with a very simple fraction!

Practice A

Reduce each fraction or product of fractions. Turn the page to check your solutions.

1. $\dfrac{12r}{20r^2}$

2. $\dfrac{24b^2 - 18b}{8b - 6}$

3. $\dfrac{8p}{11} \cdot \dfrac{22}{15p}$

4. $\dfrac{9a^2}{20b^2} \cdot \dfrac{25b}{12a}$

5. $\dfrac{6n^2 - 12n}{6n^2 - 9n}$

6. $\dfrac{2c}{8c^2d - 2d} \cdot \dfrac{12c^2d^2 - 3d^2}{3cd}$

B. Simplifying Sums of Algebraic Fractions

To evaluate a sum of algebraic fractions, such as $\dfrac{2}{9w^2} + \dfrac{5}{27w}$, we rewrite each fraction as an equivalent fraction using the LCM of the denominators as the common denominator. In this example, we use $\text{lcm}(9w^2, 27w) = 27w^2$ as the common denominator:

$$\dfrac{2}{9w^2} + \dfrac{5}{27w} = \dfrac{2}{9w^2} \cdot \dfrac{3}{3} + \dfrac{5}{27w} \cdot \dfrac{w}{w}$$

$$= \dfrac{6}{27w^2} + \dfrac{5w}{27w^2}$$

$$= \dfrac{6 + 5w}{27w^2}$$

Example 3

Simplify each sum of algebraic fractions.

1. $\dfrac{5}{3g} + \dfrac{g}{6}$

2. $\dfrac{7}{12q} - \dfrac{3}{8q^2}$

Solutions

1. First calculate the least common multiple of the denominators: $\text{lcm}(3g, 6) = 6g$

$$\frac{5}{3g} + \frac{g}{6} = \frac{2 \cdot 5}{2 \cdot 3g} + \frac{g \cdot g}{6 \cdot g}$$

To reach the common multiple of $6g$, we can multiply each fraction by the equivalent of the integer 1 so that each fraction is equivalent to its original.

$$= \frac{10}{6g} + \frac{g^2}{6g}$$

This now gives us the common denominator $6g$, which we can add together.

$$= \frac{10 + g^2}{6g}$$

We are left with one fraction, instead of two that we need to add together.

2. Again, start by calculating the least common multiple of the denominators:
$\text{lcm}(12q, 8q^2) = 24q^2$

$$\frac{7}{12q} - \frac{3}{8q^2} = \frac{7 \cdot 2q}{12q \cdot 2q} - \frac{3 \cdot 3}{3 \cdot 8q^2}$$

$$= \frac{14q}{24q^2} - \frac{9}{24q^2}$$

$$= \frac{14q - 9}{24q^2}$$

For more complicated algebraic fractions, you may need to factor each denominator in order to find the least common multiple.

Bear in mind that *after* we add the algebraic fractions that have common denominators, it may be possible to reduce them.

Example 4

Simplify each sum of algebraic fractions.

1. $\dfrac{1}{3d^2 + d} - \dfrac{1}{6d + 2}$

2. $\dfrac{2v}{9v^2 - 6v} + \dfrac{8}{15v - 10}$

Solutions

1. To help calculate $\text{lcm}(3d^2 + d, 6d + 2)$, we have to first factor each denominator separately. What we find is: $\text{lcm}(3d^2 + d, 6d + 2) = \text{lcm}(d(3d + 1), 2(3d + 1)) = 2d(3d + 1)$.

$$\frac{1}{3d^2 + d} - \frac{1}{6d + 2} = \frac{2 \cdot 1}{2 \cdot d(3d + 1)} - \frac{d \cdot 1}{d \cdot 2(3d + 1)}$$

Now we can reach the common denominator in order to find the difference between the two fractions.

$$= \frac{2 - d}{2d(3d + 1)}$$

We're left with a simplified fraction.

2. Using the same process, we can find the LCM of these fractions by factoring each denominator separately. What we find is: $\text{lcm}(9v^2 - 6v, 15v - 10) = \text{lcm}(3v(3v - 2),$ $5(3v - 2)) = 15v(3v - 2)$.

$$\frac{2v}{9v^2 - 6v} + \frac{8}{15v - 10} = \frac{5 \cdot 2v}{5 \cdot 3v(3v - 2)} + \frac{3v \cdot 8}{3v \cdot 5(3v - 2)}$$

$$= \frac{10v + 24v}{15v(3v - 2)}$$

In this problem, we can further reduce the resulting fraction by canceling the variable.

$$= \frac{34v}{15v(3v - 2)}$$

$$= \frac{34}{15(3v - 2)}$$

Practice B

Simplify each sum of algebraic fractions. Turn the page to check your solutions.

7. $\dfrac{x}{12} - \dfrac{5}{3x}$

8. $\dfrac{17}{10b} - \dfrac{5}{4b^2}$

9. $\dfrac{2}{15b^2} + \dfrac{3}{25b}$

10. $\dfrac{1}{4k + 8} + \dfrac{1}{2k^2 + 4k}$

11. $\dfrac{2n}{n + 2} - \dfrac{11}{2n^2 + 4n}$

12. $\dfrac{9}{16t + 12} + \dfrac{3t}{8t^2 + 6t}$

Exercises 4.3

For the following exercises, simplify the algebraic fraction or product of algebraic fractions.

1. $\dfrac{14m^2}{21m}$

2. $\dfrac{24q}{30q^2}$

3. $\dfrac{32}{15q} \cdot \left(-\dfrac{3q}{8}\right)$

4. $\dfrac{18b^2}{32b}$

5. $\dfrac{8}{3v} \cdot \dfrac{v}{4}$

6. $\dfrac{4v}{7} \cdot \dfrac{21}{15v}$

7. $-\dfrac{49}{15t} \cdot \dfrac{5t^2}{14}$

8. $\dfrac{y}{3} \cdot \dfrac{12}{5y}$

9. $\dfrac{10}{3t^2} \cdot \dfrac{12t}{5}$

10. $\dfrac{8x^2}{15} \cdot \dfrac{25}{16x}$

11. $\dfrac{24k + 12}{16k^2 + 8k}$

12. $-\dfrac{30x}{7} \cdot \dfrac{1}{18x^2}$

Practice A — Answers

1. $\dfrac{3}{5r}$

2. $3h$

3. $\dfrac{16}{15}$

4. $\dfrac{15a}{16b}$

5. $\dfrac{2n - 4}{2n - 3}$

6. 1

13. $\dfrac{24y - 16y^2}{21 - 14y}$

14. $\dfrac{35 + 21r}{25r + 15r^2}$

15. $\dfrac{12p^2 - 10p}{18p - 15}$

16. $\dfrac{9q - 6q^2}{10q - 15}$

17. $\dfrac{4s + 8}{-12s - 6s^2}$

18. $\dfrac{6y^2 - 2y}{5 - 15y}$

19. $-\dfrac{m + 3}{21} \cdot \left(-\dfrac{10}{5m + 15}\right) \cdot \dfrac{-3m}{4}$

20. $\dfrac{3 + 12w}{-16w^2 - 4w}$

21. $\dfrac{3}{y^2 + 2y} \cdot \dfrac{y + 2}{9}$

22. $\dfrac{5}{3r + 6} \cdot \dfrac{2r}{3} \cdot \dfrac{r + 2}{15}$

23. $\dfrac{-4v}{3} \cdot \dfrac{7}{-14v + 21} \cdot \dfrac{-2v + 3}{28}$

24. $\dfrac{3 + 4x}{10} \cdot \left(-\dfrac{5x}{6x + 8x^2}\right)$

For the following exercises, simplify the sum of algebraic fractions.

25. $\dfrac{3}{5b} - \dfrac{2b}{15}$

26. $\dfrac{n}{12} + \dfrac{7}{6n}$

27. $-\dfrac{5}{6w} + \dfrac{7w}{9}$

28. $\dfrac{4s}{3} + \dfrac{9}{5s}$

29. $\dfrac{11}{7x} + \dfrac{21}{35}$

30. $\dfrac{3m}{14} - \dfrac{4}{7m}$

31. $\dfrac{3}{7a^2} - \dfrac{16}{21a}$

32. $\dfrac{25}{24} - \dfrac{9}{10k}$

33. $\dfrac{2}{9p^2} + \dfrac{5}{12p}$

34. $\dfrac{18}{35b} - \dfrac{8}{21b^2}$

35. $\dfrac{39}{6t - 12t^2} + \dfrac{4t}{2 - 4t}$

36. $\dfrac{1}{14z} + \dfrac{6}{7z^2}$

37. $\dfrac{12}{15k^2 - 10k} + \dfrac{k}{9k - 6}$

38. $\dfrac{5x}{8z - 24} - \dfrac{12}{5z^2 - 15z}$

39. $\dfrac{4}{9w} - \dfrac{3}{2w^2 - w}$

40. $\dfrac{6k}{2s - 4s^2} + \dfrac{3}{10s - 5}$

41. $\dfrac{2m}{7m - 35} + \dfrac{1}{3m^2 - 15m}$

42. $\dfrac{9}{12q - 8q^2} - \dfrac{4}{3q}$

43. $\dfrac{3}{2 - 6q} - \dfrac{3q}{4q - 12q^2}$

44. $\dfrac{2}{3v - 1} - \dfrac{-4}{6v - 18v^2}$

45. $\dfrac{9}{30x + 50} + \dfrac{3}{6x^2 + 10x}$

46. $\dfrac{5}{4x - 2x^2} - \dfrac{5}{8 - 4x}$

7. $\dfrac{x^2 - 20}{12x}$

8. $\dfrac{34b - 25}{20b^2}$

9. $\dfrac{10 + 9b}{75b^2}$

10. $\dfrac{1}{4k}$

11. $\dfrac{4n^2 - 11}{2n(n + 2)}$

12. $\dfrac{15}{4(4t + 3)}$

4.4 Modeling with Expressions and Equations

Simone has recently taken up the hobby of rock climbing. She is visiting the Rock Boxx climbing gym in Salem, Oregon and is trying to decide which type of membership to buy. Let's look at some similar problems and see how Simone can use algebra to make the best choice.

A. Interpreting Situations with Expressions

Algebra gives us a powerful set of tools for understanding the world around us. In many cases, we can use expressions to form clear and correct ideas about relationships we observe and to effectively describe those relationships to others.

The key to writing an effective expression is to identify the unknown quantity or quantities involved in the situation and assign variables to represent them. This step is called **defining the variable(s)**. It's important to be clear and precise when we define a variable because the definition that we write lays the foundation for the rest of our work. We must also remember that the variable *must represent a number*.

Here's an example of a well written variable definition and a poorly written variable definition:

- Let p represent the number of pounds of saltwater taffy that Sue buys.

- Let p be saltwater taffy.

Can you tell which one is poorly written? The second one doesn't work, even though it's nice that we know what kind of taffy Sue likes best. The problem is that "saltwater taffy" can't be replaced by a number. In the first example, "the number of pounds of saltwater taffy" *can* be replaced by a number.

The following example highlights the difference between these two definitions.

Example 1

Larry and Sue drive out to the coast to go whale watching. While waiting for the boat to arrive, they visit the Fuddy Duddy Fudge Shop and buy saltwater taffy. Larry buys 2 pounds more than Sue buys. Write an expression to represent the amount of saltwater taffy that Larry buys.

Solution

Using our first definition of p, the expression $2 + p$ clearly represents the number of pounds of saltwater taffy that Larry buys, which is two pounds more than Sue buys. Using our second definition of p, it's unclear what p even represents, so we cannot use it effectively to write an expression.

Our next example involves the same variable definition, but the expression is a bit more complicated.

> ### Example 2
>
> The saltwater taffy costs $1.25 per pound. Sue pays with a twenty-dollar bill. Write an expression for the amount of change that she receives.
>
> #### Solution
> This expression involves two operations, subtraction and multiplication. Multiplication is used to calculate the amount of the purchase. At $1.25 per pound, p pounds of saltwater taffy amounts to a purchase of $1.25p$ dollars. We then use subtraction to calculate change. "Change" is the amount of the money in excess of the purchase. In this case, $20 was paid, and the amount of the purchase was $1.25p$, so the change would be $20 - 1.25p$.

Simone is still trying to decide which type of gym membership to buy. Because she's a student, she's eligible for the following reduced rates:

- ◆ $10 for a day pass
- ◆ $87 for a 10-visit punch card
- ◆ $49 for a one-month membership

Assuming that she uses all 10 visits on the punch card, that option becomes more economical than the day pass. The price for each visit is the total price divided by the number of visits:

$$\frac{\$87}{10 \text{ visits}} = \$8.70 \text{ per visit}$$

Over the course of 10 visits, Simone will save a total of $13 by having a punch card. The question that Simone wants to answer, though, is whether she can save even more by having a monthly membership? We can help her figure that out with the following practice exercise.

Practice A

1. Define a variable that will allow Simone to write an expression for the cost of climbing at the Rock Boxx for one month using a 10-visit punch card. When you're finished, turn the page to check your solutions.

2. Use your variable to write an expression for the monthly cost of climbing at the Rock Boxx using a 10-visit punch card.

B. Using Equations to Answer Questions

For questions that can be answered with algebra, we often need information that is not given to us directly. The information that we don't have is often implied by the information that we do have. For example, the key piece of information that Simone needs to make the most economical choice of gym membership is this: how frequently she will have to climb in order for the monthly membership to be more economical than the punch card. That piece of information is not given to Simone directly. All she has are prices of different levels of use. However, it is implied, by the pricing information, that she has to visit a certain number of times in order to make the monthly membership "worth it." She can use the expression that you wrote in Practice A to write an equation that will help her answer this question.

Let's ask the question carefully, and then let's see how asking the question is represented mathematically by an equation:

> *Question:* How many times per month must Simone use the gym in order for the punch card and the monthly membership to be the same price?

Before you write the equation, try writing the complete sentence (with words) that the answer will be embedded into. Leave a blank space for the answer. Writing out this sentence will help you define the variable well.

> *Answer:* Simone must climb _____ times per month for the punch card and the monthly membership to be the same price.

The act of writing this sentence helps us see how we must define the variable:

> Let n represent the number of times per month that Simone visits the Rock Boxx.

We don't need to include the objective of saving money in the definition of the variable because that will be represented by an equation.

We established in Practice A that the monthly cost of climbing with a punch card is $8.7n$ dollars per month. We are given the information that the cost of a monthly membership is 49 dollars. Now we simply write these two expressions into an equation:

$$8.7n = 49$$

This equation is a mathematical representation of our objective. The monthly cost of a punch card, $8.7n$, is the same as the cost of a monthly membership, 49 dollars. The solution of this equation is the number that we will insert into the blank in the sentence that we wrote:

$$8.7n = 49$$

$$n = \frac{49}{8.7}$$

$$n \approx 5.6$$

As usual, this answer needs to be interpreted carefully. Simone will not climb exactly 5.6 times in any given month because that is impossible. She either climbs, or she doesn't. She can't climb 0.6 times.

The value of this answer is that it gives Simone the information that she needs to decide what type of membership to buy. If she plans to climb 6 or more times per month, then she should buy the monthly membership. If she plans to climb 5 or fewer times per month, then she should use the punch card.

Example 3

The long-awaited Peter Courtney Minto Island Bicycle and Pedestrian Bridge, originally scheduled for completion in 2016, was finally opened in April 2017. The city of Salem, Oregon assured local residents that the project was on budget even though construction was delayed. By contractual agreement, the city is allowed to deduct $750 per day from the contract amount for each day the contractor fails to meet the 2016 deadline. It is estimated that by the time the bridge opened to the public, the city will withhold more than $200,000. How many days late must the project completion be for this to happen?

Solution

First we write the sentence into which we will embed the answer:

> The project completion must be at least _____ days late for the city to withhold more than $200,000.

Next, we define the variable:

> Let d be the number of days past deadline that the bridge was completed.

Now we write and solve an equation that represents our objective:

$$750d = 200{,}000$$
$$d = \frac{200{,}000}{750}$$
$$d \approx 267$$

Last, we fill in the blank from the first step with the solution of the equation:

> The project must be at least 267 days late for the city to withhold more than $200,000.

Practice A — Answers

1. Let n represent the number of times per month that Simone visits the Rock Boxx.

2. Using a 10-visit punch card, each visit costs $8.70. The cost for climbing at the Rock Boxx is $8.7n$ dollars per month.

Practice B

3. Simone's long rope is 4/3 the length of her short rope. If her long rope is eight meters long, then what is the length of her short rope?

4. Simone put $2.25 in the parking meter. If the parking meter rate is $1.50 per hour, how much time does she have?

5. Simone estimates that she needs to work out for 18 more minutes if she wants to exactly compensate for a 390 calorie lunch she ate earlier. Her workouts average 13 calories per minute, which means she needs to work out for 30 minutes, total. Approximately how many minutes has she already been at the gym.

6. Simone's water bottle contains 9 ounces of liquid when it is at 3/5 of its capacity. How many ounces of liquid does it contain when full?

7. Simone decides to include a climbing line item in her personal budget. This line item funds her $49 per month membership fee, but it also includes some extra money that she sets aside each month to save for a pair of climbing shoes. She wants to be able to buy a pair of La Sportive Mythos, which cost $140, at the end of 5 months. How much should Simone's monthly climbing budget be?

8. Simone spent triple the time in the gym on Friday than she spent on Tuesday, and she spent 15 minutes less time in the gym on Tuesday than she spent on Monday. If she spent 87 minutes in the gym on Friday, then how much time did she spend in the Gym on Monday?

Exercises 4.4

In the following exercises, define a variable for the unknown quantity. Use the first letter of a word related to the unknown quantity as the variable. Use a complete sentence to define the variable. The first word of that sentence should be "let." If applicable, include the units, measurement, or dimensions of the unknown quantity in your definition.

1. What is the giraffe's weight in kilograms?

2. How many minutes does it take to travel to the grocery store?

3. How many exams have been graded?

4. How many points were scored?

5. What is the length of the rectangle in miles?

6. What is the mass of the sun in kilograms?

7. What is the income of the Californian accountant in dollars per year?

8. How many games did the Oregon Ducks win in the three seasons spanning 2013–2015?

9. Larry teaches a CPR class. He gets paid an extra $20 per class to have his heart stopped so that students can practice restarting it. Larry is a nervous wreck, but he's making good money. If Larry teaches 3 classes per week, how much extra money does he earn in a year?

3. Let s be the length of Simone's short rope, in meters.

$$\frac{4}{3}s = 8$$
$$s = 6$$

Simone's short rope in 6 meters long.

4. Let h be the number of hours on the parking meter.

$$1.50\,h = 2.25$$
$$h = 1.5$$

Simone has 1.5 hours on the parking meter.

5. Let b be the number of minutes she has already been at the gym.

$$b + 18 = 30$$
$$b = 12$$

Simone has been working out for about 12 minutes.

6. Let f be the number of ounces simone's water bottle contains when full.

$$\frac{3}{5}F = 9$$
$$F = 15$$

Simone's water bottle contains 15 ounces of liquid when full.

7. Let l represent the amount of money in Simone's monthly climbing budget.

$$5(l - 49) = 140$$
$$l = 77$$

Simone should budget $77 per month for climbing.

8. Let m be the number of minutes that Simone spent at the gym on Monday.

$$(m - 15)3 = 87$$
$$m = 44$$

Simone spent 44 miutes at the gym on Monday.

10. Yesterday, Doug saw a thief that was dressed up like a janitor. The thief was in the hallway outside of Doug's classroom, stealing burned out light bulbs and replacing them with new ones. Well, Doug believes he was a thief, anyway. The recycling plant will buy used light bulbs for $1 each. New bulbs cost $14 each. How much money does this thief lose each time he steals a light bulb?

In the follow exercises, use the given information to write an expression.

11. Let *a* represent an unknown number. Write an expression for 3 plus the unknown number.

12. Let *b* represent an unknown number. Write an expression for 7 added to the unknown number.

13. Let *d* represent an unknown number. Write an expression for the sum of the unknown number and 5.

14. Let *g* represent an unknown number. Write an expression for 6 more than the unknown number.

15. Let *h* represent an unknown number. Write an expression for 4 greater than the unknown number.

16. Let *k* represent an unknown number. Write an expression for the unknown number increased by 9.

17. Let *m* represent an unknown number. Write an expression for the unknown number exceeded by three.

18. Let *n* represent an unknown number. Write an expression for 2 minus the unknown number.

19. Let *p* represent an unknown number. Write an expression for 6 less than the unknown number.

20. Let *q* represent an unknown number. Write an expression for the unknown number less 8.

21. Let *r* represent an unknown number. Write an expression for 5 reduced by the unknown number.

22. Let *s* represent an unknown number. Write an expression for 4 decreased by the unknown number.

23. Let *t* represent an unknown number. Write an expression for the unknown number subtracted from 9.

24. Let *v* represent an unknown number. Write an expression for 3 times the unknown number.

25. Let *w* represent an unknown number. Write an expression for the product of 5 and the unknown number.

26. Let *x* represent an unknown number. Write an expression for double the unknown number.

27. Let *y* represent an unknown number. Write an expression for quadruple the unknown number.

28. Let *z* represent an unknown number. Write an expression for three-fifths of the unknown number.

29. Let *a* represent an unknown number. Write an expression for the unknown number divided by 8.

30. Let *b* represent an unknown number. Write an expression for the ratio of the unknown number to 3.

31. Let *d* represent an unknown number. Write an expression for two times the quantity of 3 plus the unknown number.

32. Let *e* represent an unknown number. The quantity 4 minus the unknown number is divided by 5.

33. The Wicked Witch Bakery hired small children to clean its ovens until the bakery was shut down in 1573, when a customer overheard a witch say, "Climb all the way in and clean that spot way in the back. I promise not to shut the door." Let *Y* represent the number of years the bakery was in operation. Write an expression to represent the year that the bakery opened.

34. The recipe called for "eye of newt," so now there's a newt down by the lake with only one eye. If a witch makes this recipe n times per week, write an expression for the number of newts lacking depth perception at the end of three weeks.

35. Jane is a psychic, but she only makes predictions about the outcomes of things that have already happened. Jane isn't a very good psychic. Even so, she predicts that the local soccer team will win a game yesterday by 3 points. If the losing team scored P points, write an expression for the winning team's score.

36. If Sue can work 3 algebra problems every 10 minutes, write an expression for the number of problems she works if she stays awake doing algebra homework for h hours on the night before a test.

For each of the following exercises, define a variable to represent the unknown quantity. Then write an equation that models the situation and solve the equation.

37. Twenty-eight less than an unknown number is equal to 53. What is the unknown number?

38. Five increased by an unknown number is equal to 40. What is the unknown number?

39. The product of 5 and an unknown number is equal to 40. What is the unknown number?

40. The ratio of an unknown number to 17 is equal to 23. What is the unknown number?

41. Twenty-two divided by the quantity 4 plus an unknown number is equal to 2. What is the unknown number?

42. Fifty-five divided by the quantity 13 minus an unknown number is equal to 11. What is the unknown number?

43. Twenty-eight is decreased by 15 times an unknown number, and the result is −17. What is the unknown number?

44. The product of 21 and an unknown number is 17 greater than 17. What is the unknown number?

45. Negative 99 is equal to triple the quantity 23 plus an unknown number. What is the unknown number?

46. Six is equal to 24 plus double an unknown number. The result is divided by 3. That result is 6. What is the unknown number?

47. Two multiplied by an unknown number is subtracted from 13. The result is split into 7 equal pieces. Each piece has the value 5. What is the unknown number?

48. Three times an unknown number is subtracted from 14. The result is split into 5 pieces. Each piece has the value −2. What is the unknown number?

49. One third less than an unknown number is equal to five thirds. What is the unknown number?

50. An unknown number is increased by four ninths. The result is seven ninths. What is the unknown number?

51. Four fifths is equal to eight fifteenths of an unknown number. What is the unknown number?

52. The opposite of 27 halves multiplied by an unknown number is equal to 9. What is the unknown number?

53. Pam and Pat teach algebra at the local community college. Pam cares about her students, but Pat tricks his students into thinking that he cares about them by buying donuts. Seventeen more students pass Pam's class than Pat's class. Thirty students pass Pam's class. How many students pass Pat's class?

54. Rick plans to use dynamite to weed his lawn. If there are 45 dandelions in his lawn, and if Rick can destroy 5 dandelions with each stick of dynamite, then how many sticks of dynamite will Rick need for his task?

55. Jim returned his calculus text to the bookstore because it was upside down. "This book is defective!" he shouted. The clerk took the book back to the stockroom and returned with it right side up. "Here is your replacement," she said, "but there will be a $8 restocking fee." If the clerk has $3 left from the restocking fee after buying a caramel frappe, how much does a caramel frappe cost?

56. Larry has invented a new treatment for tennis elbow. First, you fill a paper cup with water, and then put the cup of water in the freezer. After the water is frozen, you roll up your sleeve and cut your arm off. According to Larry, cutting your arm off hurts less than rubbing the ice on your elbow. If Larry gets 0.4 cents per viewer of an online video showing his new treatment, how many viewers does he need to earn $1000?

57. Stan has a pile of bricks that weighs a total of 44 pounds. Unfortunately, Stan is so strong that he can only lift piles that weigh 80 pounds or more. If a brick weighs 3 pounds, then how many bricks must Stan add to the pile before he can lift it?

58. The local nursery will deliver 3 cubic yards of bark for $120. One cubic yard is equivalent to 13.5 bags of bark. If bags cost $4 each, how many bags of bark would cost the same as a delivery of three cubic yards?

59. If the energy operating cost of your very old refrigerator is $117 per year and the cost of operating a new refrigerator is $40 per year, how much can you spend on a new refrigerator if you want the cost to break even in 10 years?

60. A gas tank holds 14 gallons when it is at $\frac{7}{9}$ of its capacity. What is its capacity?

61. Walking 63 feet is $\frac{7}{8}$ of the length of a walkway. What is the total length of the walkway?

62. A 60-watt bulb costs $1 and lasts for 1000 hours. A replacement is a 9-watt LED bulb that costs $3 and lasts for 15,000 hours. If the lifetime operating energy cost for a 60W bulb is $6.30, but for an LED bulb is $14.18, how much money would the LED bulb save over its lifetime?

63. Bonnie wants to start selling her widgets online. An e-commerce company will host her store on their servers for $28/month. She can also buy a one-time license for their software for $495. In that case, she will also have to pay $8/month to a web site hosting company and install the software there herself. What is the break-even point?

64. When Joe was a baby and his mother spooned food into his mouth, she used to say, "Choo-choo! Chugga chugga chugga! Choo-choo! Open wide!" Fifty years later, Joe can't remember this childhood trauma at all. However, when he hears a train, he can't resist jumping onto the tracks with his mouth open. If it is now 8:00 am, and if the train is scheduled to pass by at 11:00 am, how many hours does Joe have left to live?

65. Ben is a teenager. His mother entered him in a toddler race at the local fair. She said that it would give him confidence if he could win a race once in a while. If the cutoff age for the race is two years old, and if Ben is 14 years old, how many years late to the race is he?

4.5 Solving Equations

Early in this book, we used fact-families to solve equations. As we transitioned from whole numbers to integers and then to rational numbers, we adopted more sophisticated techniques for solving equations by using the addition principle and the multiplication principle.

In this section, we add two more layers of complexity to types of equations that we can solve:

- ◆ equations in which the variable is present on both sides of the equation

- ◆ equations in which the expressions on one or both sides are not simplified

A. Reviewing Key Algebraic Definitions

Let's review some of the definitions and principles we need to use the language of algebra:

An **equation** is a statement that two expressions evaluate to the same number. These two expressions are written on opposite sides of an equal sign. Each expression may be arithmetic, which has no variables included, or algebraic, which involves at least one variable.

A **solution** of an equation with one variable is a value of that variable that makes the equation a true statement.

Two **equations are equivalent** if every solution of the first equation is also a solution of the second equation and if every solution of the second equation is also a solution of the first equation.

Two **algebraic expressions are equivalent** if each expression evaluates to the same number as the other under any substitution (of a number for the variable).

The **addition principle of equations** states that if A and B are any mathematical expressions, either algebraic or arithmetic, and c is any rational number, then the equations $A = B$ and $A + c = B + c$ are equivalent.

The **multiplication principle of equations** states that if A and B are any mathematical expressions, either algebraic or arithmetic, and c is any rational number, then $A = B$ and $Ac = Bc$ are equivalent as long as $c \neq 0$.

The **distributive property** states that if a, b, and c are any three numbers, then the expressions $a(b + c)$ and $ab + ac$ are equivalent.

The **equivalent expressions principle** states that if A and B are equivalent expressions and C is any expression, then the equations $A = C$ and $B = C$ are equivalent.

The **inverse properties of addition and multiplication** state that if r is any rational number then:

- ◆ There exists a unique rational number $-r$ such that $r + (-r) = -r + r = 0$

- ◆ If $r \neq 0$ there exists a unique rational number $\frac{1}{r}$ such that $r \cdot \frac{1}{r} = \frac{1}{r} \cdot r = 1$

We call $-r$ the **additive inverse** of r or, more informally, the **opposite** of r. We call $\frac{1}{r}$ the **multiplicative inverse** of r, or more informally, the **reciprocal** of r.

Practice A

Identify the above principle or property involved in performing the following steps:

1.
$$\frac{2}{3}x = \frac{5}{7}$$
$$\left(\frac{3}{2}\right)\frac{2}{3}x = \left(\frac{5}{7}\right)\frac{3}{2}$$

2.
$$\frac{7}{4}\left(\frac{4}{7}\right)x = 13$$
$$1 \cdot x = 13$$

3.
$$21 = 7x + 4 + (-4)$$
$$21 = 7x + 0$$

4.
$$57 = 2x + 3$$
$$57 + (-3) = 2x + 3 + (-3)$$
$$54 = 2x$$

5.
$$7x = 2(3x + 4)$$
$$7x = 6x + 8$$

6.
$$4x = 18$$
$$\left(\frac{1}{4}\right)4x = 18\left(\frac{1}{4}\right)$$

B. Variable on Both Sides of an Equation

Now we can get to work with more complex equations. Here's one type of equation that we haven't encountered yet:
$$4x + 7 = 23 - 2x$$

This is the first equation we've seen in which the left-hand side and right-hand side are both *algebraic* expressions. To solve this equation, we must find an equivalent equation with the form $x =$ (some number) so that at some point in the solution process, one side of the equation will become an *arithmetic* expression. We will use the addition principle to do this:

$4x + 7 = 23 - 2x$	This is the equation we wish to solve.
$4x + 7 + 2x = 23 - 2x + 2x$	Apply the addition principle.
$6x + 7 = 23$	Simplify both sides by combining like terms.
$6x + 7 + (-7) = 23 + (-7)$	Apply the addition principle.
$6x = 16$	Now evaluate sums.
$\frac{1}{6} \cdot 6x = \frac{1}{6} \cdot 16$	Apply the multiplication principle.
$x = \frac{8}{3}$	Evaluate products.

Now we check our work.
$$4 \cdot \frac{8}{3} + 7 = 23 - 2 \cdot \frac{8}{3}$$
$$\frac{32}{3} + \frac{21}{3} = \frac{69}{3} - \frac{16}{3}$$
$$\frac{53}{3} = \frac{53}{3} \qquad \text{It checks out!}$$

Example 1

Solve each equation

1. $5x - 12 = 8x + 15$

2. $\frac{3n}{8} - \frac{7}{12} = \frac{9}{4} - \frac{5n}{6}$

3. $2v - 5 = 2v + 12$

4. $7q + 4 = 5q + 4$

Solutions

As you read these solutions, take a close look and see if you can identify which equation solving principle is used in each step.

1. $5x - 12 = 8x + 15$

$$5x - 12 = 8x + 15$$
$$-5x + 5x - 12 = -5x + 8x + 15$$
$$-12 = 3x + 15$$
$$-12 + (-15) = 3x + 15 + (-15)$$
$$-27 = 3x$$
$$-9 = x$$

Check your work:

$$5(-9) - 12 = 8(-9) + 15$$
$$-45 - 12 = -72 + 15$$
$$-57 = -57 \checkmark \qquad \text{It checks!}$$

2. $\frac{3n}{8} - \frac{7}{12} = \frac{9}{4} - \frac{5n}{6}$

$$\frac{3n}{8} - \frac{7}{12} = \frac{9}{4} - \frac{5n}{6}$$
$$\left(\frac{3n}{8} - \frac{7}{12}\right) \cdot 24 = \left(\frac{9}{4} - \frac{5n}{6}\right) \cdot 24$$
$$\frac{3n}{8} \cdot 24 - \frac{7}{12} \cdot 24 = \frac{9}{4} \cdot 24 - \frac{5n}{6} \cdot 24$$
$$9n - 14 = 54 - 20n$$
$$29n = 68$$
$$n = \frac{68}{29}$$

Check your work:

$$\frac{3}{8} \cdot \frac{68}{29} - \frac{7}{12} = \frac{9}{4} - \frac{5}{6} \cdot \frac{68}{29}$$

$$\frac{3}{48} \cdot \frac{3468}{29} - \frac{7}{12} = \frac{9}{4} - \frac{5}{36} \cdot \frac{3468}{29}$$

$$\frac{102}{116} \cdot \frac{3}{3} - \frac{7}{12} \cdot \frac{29}{29} = \frac{9}{4} \cdot \frac{87}{87} - \frac{170}{87} \cdot \frac{4}{4}$$

$$\frac{306}{348} - \frac{203}{348} = \frac{783}{348} - \frac{680}{348}$$

$$\frac{103}{348} = \frac{103}{348} \checkmark \qquad\qquad \text{It checks!}$$

3. $2v - 5 = 2v + 12$

This equation has *no solution* because it is equivalent to a false statement.

$$2v - 5 = 2v + 12$$

$$-2v + 2v - 5 = -2v + 2v + 12$$

$$-5 = 12$$

4. $7q + 4 = 5q + 4$

$$7q + 4 = 5q + 4$$

$$-5q + 7q + 4 = -5q + 5q + 4$$

$$2q + 4 = 4$$

$$2q + 4 + (-4) = 4 + (-4)$$

$$2q = 0$$

$$\frac{1}{2} \cdot 2q = \frac{1}{2} \cdot 2q$$

$$q = 0$$

Check your work:

$$7 \cdot 0 + 4 = 5 \cdot 0 + 4$$

$$0 + 4 = 0 + 4$$

$$4 = 4 \checkmark \qquad\qquad \text{It checks!}$$

Practice A — Answers

1. Multiplication principle of equations
2. Inverse property of multiplication
3. Inverse property of addition
4. Addition principle of equations
5. Distributive property
6. Multiplication principle of equations

In second problem above, there's an interesting use of the multiplication principle in the first step. We multiplied both sides of the equation by the least common multiple of the denominators of *all of the fractions* in the equation, or lcm(8, 12, 4, 6) = 24. This produces an equivalent equation in the fourth line that has no fractions. This useful trick is called **clearing the denominators**.

Practice B

Solve each of the following equations. Turn the page to check your answers.

7. $3x - 4 = 2x + 5$

8. $6p + 11 = 4p - 3$

9. $-7k + 4 = 2 + 5k$

10. $-4r + 2 = 8 - 4r$

11. $\frac{2}{3}y - \frac{5}{4} = \frac{3}{4} + \frac{1}{3}y$

12. $\frac{7a}{18} - \frac{5}{12} = \frac{5a}{9} + \frac{11}{6}$

C. Expressions That Can Be Simplified

Consider the equation $4(2x - 5) = 28$. We can solve this equation using the addition and multiplication principles to undo operations in reverse order, but simplifying the left-hand side first gives us a more streamlined solution. By the distributive property, $4(2x - 5)$ and $8x - 20$ are equivalent expressions. By the equivalent expressions property, $4(2x - 5) = 28$ and $8x - 20 = 28$ are equivalent equations, but $8x - 20 = 28$ is easier to solve:

$$8x - 20 + 20 = 28 + 20$$

$$8x = 48$$

$$\frac{1}{8} \cdot 8x = \frac{1}{8} \cdot 48$$

$$x = 6$$

Example 2

Solve each of the following equations. Make sure both sides are completely simplified first.

1. $5y - 8 + 13y = 7 + 11y + 18$

2. $5z - (16 - 2z) = 12$

3. $2t + 5(3 - 4t) = 9 - 4(7t - 3)$

Solutions

1. $5y - 8 + 13y = 7 + 11y + 18$
    First simplify both sides by combining like terms, then solve:

$$5y - 8 + 13y = 7 + 11y + 18$$

$$18y - 8 = 25 + 11y$$

$$7y = 33$$

$$y = \frac{33}{7}$$

Check your work:

$$5 \cdot \frac{33}{7} - 8 + 13 \cdot \frac{33}{7} = 7 + 11 \cdot \frac{33}{7} + 18$$

$$\frac{165}{7} - 8 \cdot \frac{7}{7} + \frac{429}{7} = 7 \cdot \frac{7}{7} + \frac{363}{7} + 18 \cdot \frac{7}{7}$$

$$\frac{165}{7} - \frac{56}{7} + \frac{429}{7} = \frac{49}{7} + \frac{363}{7} + \frac{126}{7}$$

$$\frac{538}{7} = \frac{538}{7} \checkmark \qquad\qquad \text{It checks!}$$

2. $5z - (16 - 2z) = 12$

This one is tricky! First rewrite the subtraction as an equivalent addition, then simplify using the distributive property, then proceed to solve:

$$5z - (16 - 2z) = 12$$

$$5z + (-1)(16 - 2z) = 12$$

$$5z + (-1)16 - (-1)2z = 12$$

$$5z - 16 + 2z = 12$$

$$7z - 16 = 12$$

$$7z = 28$$

$$z = 4$$

Check your work:

$$5 \cdot 4 - (16 - 2 \cdot 4) = 12$$

$$5 \cdot 4 - (16 - 8) = 12$$

$$5 \cdot 4 - 8 = 12$$

$$20 - 8 = 12$$

$$12 = 12 \checkmark \qquad\qquad \text{It checks!}$$

3. $2t + 5(3 - 4t) = 9 - 4(7t - 3)$

This one is the full meal deal! In fact, you won't see any equations more complicated than this one in this whole book. To solve it, first use the distributive property to simplify both sides. Be sure to change the subtraction to an equivalent addition on the right-hand side.

$$2t + 5(3 - 4t) = 9 - 4(7t - 3)$$

$$2t + 5(3 - 4t) = 9 + (-4)(7t - 3)$$

$$2t + 5 \cdot 3 - 5 \cdot 4t = 9 + (-4)7t - (-4)3$$

$$15 - 18t = 21 - 28t$$

$$10t = 6$$

$$t = \frac{3}{5}$$

Check your work:

$$2 \cdot \frac{3}{5} + 5\left(3 - 4 \cdot \frac{3}{5}\right) = 9 - 4\left(7 \cdot \frac{3}{5} - 3\right)$$

$$2 \cdot \frac{3}{5} + 5\left(3 \cdot \frac{5}{5} - \frac{12}{5}\right) = 9 - 4\left(\frac{21}{5} - 3 \cdot \frac{5}{5}\right)$$

$$2 \cdot \frac{3}{5} + 5\left(\frac{15}{5} - \frac{12}{5}\right) = 9 - 4\left(\frac{21}{5} - \frac{15}{5}\right)$$

$$2 \cdot \frac{3}{5} + 5 \cdot \frac{3}{5} = 9 \cdot \frac{5}{5} - 4 \cdot \frac{6}{5}$$

$$\frac{6}{5} + \frac{15}{5} = \frac{45}{5} - \frac{24}{5}$$

$$\frac{21}{5} = \frac{21}{5} \checkmark \qquad\qquad \text{It checks!}$$

Practice C

Solve each equation. Then check your answers below.

13. $2a + 3a = 45$

14. $3(4g - 5) = 62 + g$

15. $\frac{2x}{3} + \frac{3x}{4} = \frac{5}{12}$

16. $7 - (3m - 12) = -8$

17. $7u - 12 + 6u = 11 - 4u - 23$

18. $4k - 7(3k - 2) = 20 + 4(3k - 6)$

Practice B — Answers

7. $x = 9$

8. $p = -7$

9. $k = \frac{1}{6}$

10. No solution

11. $y = 6$

12. $a = -\frac{27}{2}$

Exercises 4.5

Solve each of the following equations. Be sure to check your solutions, too.

1. $4k = 3k + 5$

2. $7y = 6y + 3$

3. $9g - 3 = 5g$

4. $5a = 8a - 4$

5. $7b - 20 = 13b + 31$

6. $2 + 8w = 15w - 26$

7. $43 - 8t = 12t - 37$

8. $33 - 9v = -5v + 21$

9. $7n = 3 + 12n$

10. $13p - 26 = 15p$

11. $\frac{2d}{3} + \frac{1}{2} = \frac{5d}{6}$

12. $\frac{1}{2} - \frac{4t}{5} = \frac{3t}{10}$

13. $\frac{8r}{15} - \frac{11}{6} = \frac{7r}{10} + \frac{1}{2}$

14. $\frac{3m}{8} - \frac{23}{12} = \frac{11m}{24} - \frac{5}{4}$

15. $2g = \frac{1}{6}g - \frac{11}{18}$

16. $\frac{1}{2}a = \frac{3}{2}a - 7$

17. $\frac{7}{12}x + \frac{3}{4} = \frac{17}{24}x + \frac{2}{3}$

18. $\frac{28}{15}m + \frac{12}{7} = \frac{11}{5}m + \frac{37}{42}$

19. $\frac{8}{5}n + 2 = 4n$

20. $3p + \frac{2}{3} = \frac{2}{3}p$

21. $35 - 8n - 16 = -n - 23$

22. $11 - 13z = 20 - 1 - 14z$

23. $13 - x + 5 = 3x - 7x$

24. $20 + 20w = -10w - 15 + 30w$

25. $7l - 15 + 3l = 26 + 10l - 8$

26. $8v + 11 - 3v = 23 - v + 5$

27. $9r + 5 - 4r = 15 + 5r$

28. $7d + d = 23d + 17 - 15d$

29. $43 + 18g - 23 = 6g - 35 + g$

30. $14 - 9x + 15 = 7x + 29 - 3x$

31. $4 + 3t - 3 - 4t = 4t + 3 - 3t$

32. $11 + y - 1 = 11y - 11 + y + 1$

33. $5p - 6 + 13p = 19 + 8p - 25$

34. $5k + 12 - 2k = 42 + 3k - 13$

35. $2(3q - 4) = 5q - 6$

36. $b = 2b - 3(4b - 5)$

37. $3y - 2(5y - 7) = 10 + y$

38. $5h - 19 = 3 - 6(3h - 4)$

39. $4(5v + 2) = 7v - 3(2 + 5v)$

40. $3(2 - m) = m - 2(3 - 4m)$

41. $5j - 3(7 - 4j) = 18 + 5(3 - 2j)$

42. $7n - 3(2 + 5n) = 9 - 4(5n + 2)$

43. $\frac{3}{5} = \left(\frac{4}{5} + \frac{5}{z}\right)2$

44. $\frac{1}{2s} = \left(\frac{3}{4} + \frac{2}{3s}\right)\frac{2}{3}$

45. $\left(\frac{5}{6k} - \frac{1}{3}\right)3 = \frac{3}{10k}$

46. $2 - \left(-\frac{1}{2}g\right) = -\left(\frac{5}{2}g + \frac{1}{4}\right)$

47. $-\left(\frac{5}{4}x + \frac{1}{3}\right) = \frac{5}{4} - \left(\frac{x}{2} + \frac{2}{3}\right)$

48. $\left(\frac{6}{11} + \frac{5}{n}\right)\frac{2}{5} = \frac{8}{15}$

50. $-\frac{13}{2} + \frac{2}{3}y = \frac{5}{8} + \frac{18}{27}y$

49. $\frac{6c}{5} + \frac{5}{4} = \frac{9}{2} + \frac{18c}{15}$

CHAPTER 5
Graphs and Tables

What's a Number?

We use numbers, in the form of data, to tell stories. This can be as simple as describing the size or amount of something we encountered, like when you tell a friend you ate 6 slices of pizza for lunch and you're still hungry. Data are also used to tell complex stories of interwoven meaning, like a census report that is used to determine how social services dollars should be divided and spent across categories of demographic information.

The value of numbers, in the second sense, is that they can be used to generalize, record, and predict outcomes based on information. In areas like government, advertising, and social services, those numbers can be used to help people make big decisions. If you tracked the number of pizza slices you ate for a year, you could probably use that data to make some important decisions, too.

In this chapter, we study tools that allow us to communicate and interpret these stories. The tools in this chapter include graphs — visual representations of data — and tables — complex lists that are ordered based on their horizontal and vertical alignment of similar information.

5.1 Interpreting Graphical Representations of Data

In this section, we look at some common ways that data can be summarized and organized into a graphical display. Specifically, we will look at bar graphs and pie graphs.

A. Bar Graphs

A bar graph is used to visually show data values that have been sorted into categories. It consists of several "bars" that start from a zero point and extend either upward from a horizontal line or to the right from a vertical line. Each bar represents a category of information, and each entry in the data set fits into one of those categories. The height or length of each bar represents the number of data entries that fit the category for that bar.

For example, in 2016, the Silverton Hospital in Silverton, OR recorded 1,467 live births. In the hospital records, there are several pieces of data relating to each birth, such as length, weight, and sex of the baby, date and time of delivery, number of hours of labor, vital statistics of both mother and baby, etc. Each one of these bits of information is a data value. Consider the date of delivery, the baby's birthday. If we were to collect this data value for all 1,467 births, we would have a list of 1,467 dates. While this unorganized list of dates represents a complete data set, it may not be very useful. In order to easily make sense of it, we may choose to sort the birth dates into categories. We could count the number of births recorded in each month. The bar graph in Figure 1 is a visual representation of this way of summarizing and organizing information.

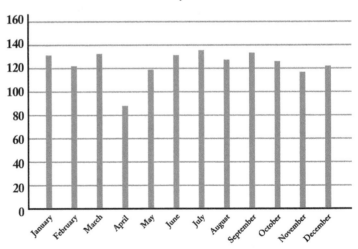

Figure 1

Example 1

Use the bar graph in Figure 1 to answer the following questions.

1. About how many babies were born in Silverton Hospital in February of 2016?

2. In which month did the fewest births occur? About how many?

3. In which month did the most births occur? About how many?

4. About how many Spring (i.e. March, April and May) births were there?

5. About how many more births were there in June than in April?

Solutions

1. About 120 babies were born in Silverton Gospital in February of 2016.

2. There were fewer births in April, about 85, than in any other month.

3. There were more births in July, about 135, than in any other month.

4. About 333 babies were born in the Spring.

5. About 45 more babies were born in June than in April.

Practice A

Answer the questions related to the following vertical bar graph. Turn the page to check your solutions.

1. What are the units of measurement for the numbers along the vertical axis?

2. What was the average monthly data usage in 2017?

3. In which year was the average monthly usage 1.7 Gigabytes per month?

4. How much did the average monthly data usage increase from 2016 to 2017?

5. How much is the average monthly data usage expected to increase from 2020 to 2021?

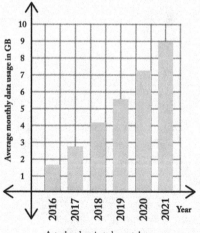

Actual and projected smartphone monthly data usage (Source: Statista)

B. Pie Graphs

A pie graph also shows data sorted into categories. The graph consists of a circle divided into several wedges. Each wedge corresponds to one of the categories, and the size of each wedge represents the number of data values in each category. A pie chart is especially useful when there are a limited number of categories and you wish to emphasize the *percentage* of data values that fall into each category.

For example, the Peter Courtney Minto Island Bicycle and Pedestrian Bridge was recently completed in Salem, Oregon. This project was funded by a number of local and state government grants along with considerable community fund raising. Funding for the bridge — which unites three parks, about 20 miles of multi-modal trails, and a parkland larger than New York's Central Park — includes $2,500,000 from Oregon Dept. of Transportation (ODOT), $1,581,000 from the State Transportation Improvement Program (STIP), $750,000 from Oregon Dept. of Parks and Recreation, and $6,200,000 from Salem's Urban Renewal Area money. The pie graph in Figure 2 shows the rough percentage of total funds that come from these various sources.

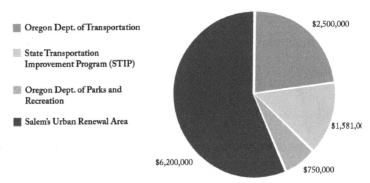

Peter Courtney Minto Island Bicycle and Pedestrian Bridge Funding Sources

- Oregon Dept. of Transportation
- State Transportation Improvement Program (STIP)
- Oregon Dept. of Parks and Recreation
- Salem's Urban Renewal Area

$2,500,000

$1,581,00

$6,200,000

$750,000

Figure 2

> ### Example 2
>
> Use the pie graph in Figure 2 to answer the following questions.
>
> 1. Did the majority of funding come from local sources of from state sources?
>
> 2. The Oregon Department of Parks and Recreation contributed less funding than any of the other funding sources. The State Transportation Improvement Program contributed about (Choose one):
>
> ◆ the same amount?
>
> ◆ twice as much?
>
> ◆ three times as much?
>
> ◆ half as much?

Solutions

1. More than half of the funding came from the Salem Urban Renewal fund, a local source.

2. The State Transportation Improvement Program contributed about twice as much money as the Oregon Department of Parks and Recreation.

Practice B

Answer the questions related to the following pie graph. Turn the page to check your solutions.

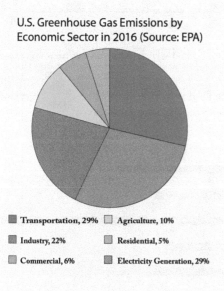

U.S. Greenhouse Gas Emissions by Economic Sector in 2016 (Source: EPA)

■ Transportation, 29%   □ Agriculture, 10%

■ Industry, 22%   ■ Residential, 5%

□ Commercial, 6%   ■ Electricity Generation, 29%

6. Which sector accounts for 6% of the total?

7. What percentage did agriculture contribute?

8. Together, commercial and residential contributed what percentage of the total?

9. Which sector accounted for most of the U.S. greenhouse gas emissions in 2016? What percentage did this sector contribute?

10. Adding all of the percentages, what is the result? Why?

11. Sectors other than electricity generation are responsible for what percentage?

Practice A — Answers

1. Gigabytes per month, which may also be written as $\frac{GB}{Mo}$

2. About 2.8 Gigabytes per month.

3. The year 2016.

4. About 1.1 Gigabytes per month.

5. About 1.7 Gigabytes per month.

Exercises 5.1

The average number of eggs per bird nest in a park is shown in the following bar graph for various years. Use the bar graph to answer the following questions.

1. About how many eggs per nest were there on average in 2004?

2. In which year was there an average of 4.3 eggs per nest?

3. In which year was there the highest average? About how many?

4. In which year were there the fewest eggs per nest on average? About how many?

5. About how many more eggs per nest on average were there in 2003 than in 2001?

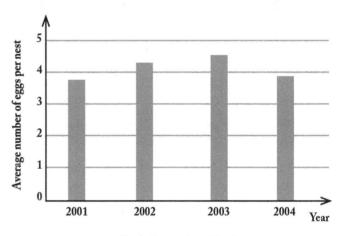

Park Eggs Per Nest

Introductory algebra students were asked what their favorite food was. The six most common answers are shown in the following bar graph. Use the bar graph to answer the following questions

6. How many students said that Sushi was their favorite food?

7. Which food was mentioned most often? How many students said that it was their favorite food?

8. Of these foods, which was mentioned least often?

9. How many students liked chicken or steak the most?

10. How many more students liked chicken best than liked steak best?

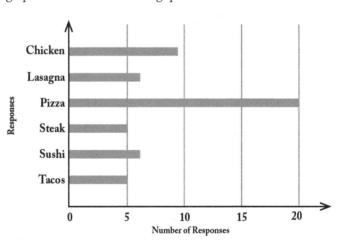

Introductory Algebra Student Favorite Foods

Grades for an introductory algebra class of thirty students were distributed as shown in the following pie graph. Use the pie graph to answer the following questions.

11. What grade corresponds to 33% of the total grades given?

12. What percentage of students earned A's?

13. What percentage of students did not get A's?

14. How many students earned B's?

15. What percentage passed with a C or better?

Course Grades

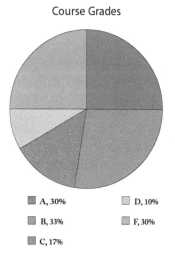

A, 30%   D, 10%

B, 33%   F, 30%

C, 17%

Answer the questions related to the following pie graph.

16. Which gas accounts for 6% of the total?

17. What percentage did methane contribute?

18. Together, Nitrous Oxide and Methane accounted for what percentage of the total?

19. Which gas accounted for the most of the U.S. Greenhouse Gas Emissions in 2016? What percentage did this gas contribute?

20. Gases other than Carbon Dioxide are responsible for what percentage?

21. Adding all of the percentages, what is the result? Why?

U.S. Greenhouse Gas Emissions in 2016 (Source: EPA)

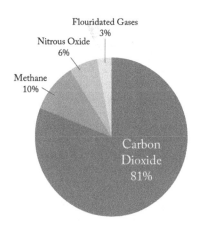

Flouridated Gases 3%

Nitrous Oxide 6%

Methane 10%

Carbon Dioxide 81%

6. The commercial sector

7. 10%

8. 11%

9. Transportation and Electricity Generation tied for the most greenhouse gas emissions in 2016, each contributing 29% of the total.

10. 101%. While all of something is 100% of it, the percentages may have been rounded, so they may not add to 100%. In fact, keeping another digit of precision for each percentage, we have Transportation 28.7%, Electricity Generation 28.6%, Industry 21.7%, Agriculture 9.5%, Commercial 6.4%, and Residential 5.1%, which add to 100.0%.

11. 71%

5.2 Mean, Median, and Mode

In this section, we study three ways to summarize a data set. The goal of these summaries is to reduce a data set into a single number. This gives you a snap-shot of the data set. For example, on a recent algebra midterm test, the student scores were:

$$85, 75, 73, 86, 60, 83, 87, 71, 64, 81, 56, 64, 88, 71, 58, 92, 58, 69, 96, 84, 71, 61, 82, 79$$

Most students like to know how the class as a whole performed on a test. Let's look at three different ways to answer that question.

A. Mean is the Average

The arithmetic **mean** of a set of numbers is the quotient of the *sum* of the numbers and the *number* of numbers in set. In the example involving midterm exam scores, there are 24 midterm scores listed. Their sum is 1794, so the mean test score is $\frac{1794}{24}$ = 74.75. The arithmetic mean of a set of numbers is often called the **average** of the set of numbers. Using this term, we would say the **average** test score was 74.75.

Example 1

The following table lists the heights of each player on the 2016 OSU women's basketball team. Find the mean height.

Player	Height (in.)
Mikayla Pivec	70
Madison Washington	73
Breanna Brown	75
Katie McWilliams	74
Gabriella Hanson	71
Kolbie Orum	75

Player	Height (in.)
Janessa Thropay	74
Marie Gulich	77
Kat Tudor	72
Sydney Wiese	73
Taylor Kalmer	68
Tarea Green	76

Solution

There are 12 players listed. Their heights add up to 878 inches, so the mean height is $\frac{878}{12} \approx 73.2$ in.

Sometimes there are too many numbers in a data set to list them all. For example, consider listing the ages of every resident of the State of Oregon. That would be a very long list! Instead of writing this out as a list of numbers, we might choose to display this data set as a **histogram**, which is a bar graph that displays a data set sorted into *bins*. Each bin is a range of values, and every number in the data set belongs in exactly one of the bins. The height of each bar is the number of data values that is sorted into the corresponding bin.

The histogram in Figure 1 from the U.S. Census Bureau shows the age distribution of Oregon residents. We can visually approximate the length of each bar, but the data labels give us the exact lengths.

For example, without the data labels, we would only be able to tell that there are approximately 300,000 Oregonians between the ages of 55 and 64, but the data label tells us that there are exactly 304,388 Oregonians between the ages of 55 and 64. Some histograms provide data labels, and others do not.

Because we don't have the actual list of ages, we can't directly calculate the mean of this data set. We can, however, use the histogram to approximate the mean. By adding up the heights of the bars, we get the number of people represented in this data set, which is 3,421,399.

To approximate the **sum** of the ages in the data set, we represent each bin with a single age. We use the ages 2.5, 10, 20, 30, ... , 90 to represent these age ranges. With the exception of the last bin, this is the age that is in the middle of the range of ages represented by that bin. We calculate the mean age using the false assumption that the age of every Oregonian is one of these representative values. We assume, for

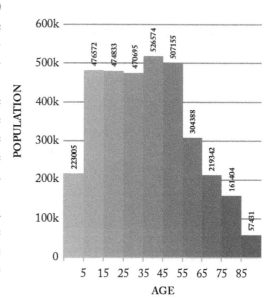

Figure 1

example, that every Oregonian between 55 and 64 years of age is *exactly* 60 years old. This assumption allows us to add up the ages by multiplying each representative age by the corresponding bar height and then adding those products:

			Running total
$2.5 \cdot 223{,}005$	$=$	$557{,}512.5$	$557{,}512.5$
$10 \cdot 476{,}572$	$=$	$4{,}765{,}720$	$5{,}323{,}232.5$
$20 \cdot 474{,}833$	$=$	$9{,}496{,}660$	$14{,}819{,}892.5$
$30 \cdot 470{,}695$	$=$	$14{,}120{,}850$	$28{,}940{,}742.5$
$40 \cdot 526{,}574$	$=$	$21{,}062{,}960$	$50{,}003{,}702.5$
$50 \cdot 507{,}155$	$=$	$25{,}357{,}750$	$75{,}361{,}452.5$
$60 \cdot 304{,}388$	$=$	$18{,}263{,}280$	$93{,}624{,}732.5$
$70 \cdot 219{,}342$	$=$	$15{,}353{,}940$	$108{,}978{,}672.5$
$80 \cdot 161{,}404$	$=$	$12{,}912{,}320$	$121{,}890{,}992.5$
$90 \cdot 57{,}431$	$=$	$5{,}168{,}790$	$127{,}059{,}782.5$

So, the approximate sum of ages is 127,059,782.5

We now calculate the approximate mean age by evaluating the quotient of the *approximate* sum of ages and the *number* of ages: The average Oregonian is $\dfrac{127{,}059{,}782.5}{3{,}421{,}399} \approx 37.1$ years old.

Practice A

Calculate or approximate the mean of each data set. Turn the page and check your solutions.

1. $\{4, 5, 8, 7, 4\}$

2. $\{3, 3, 2, 6, 9, 7\}$

3. $\{24, 59, 31, 8, 125, 47, 92, 23\}$

4.

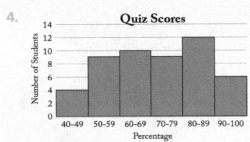

5.

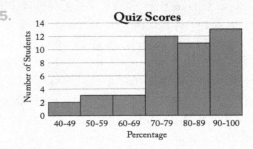

6.

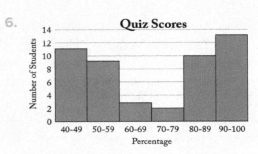

B. Median is the Middle

The **median** of a data set divides the data set into the upper half and the lower half. At least half of the numbers in the data set must be greater than or equal to the median, and at least half of the numbers in the data set must be less than or equal to the median. Here is how we calculate the median of a data set.

First, sort the data values into either increasing or decreasing order. This is very important. The numbers must be in order for the middle value to be the median.

Next, we find the middle of the list. If there is an odd number of numbers in the set, then the median is the number in the middle:

$$33, 56, 78, 79, 83$$

If there is an even number of numbers in the data set, then the median is the average of the two middle numbers. Here are the midterm exam scores from the beginning of the section written in decreasing order:

$$96, 92, 88, 87, 86, 85, 84, 83, 82, 81, 79, 75, 73, 71, 71, 71, 69, 64, 64, 61, 60, 58, 58, 56$$

There are 24 numbers, so the twelfth and the thirteenth numbers, which are highlighted here, are in the middle. The median is the average of these two numbers:

$$\frac{75 + 73}{2} = 74$$

There are exactly 12 test scores greater than 74 and 12 test scores less than 74.

Example 2

Oregon's population is divided fairly evenly between its five congressional districts, but its landmass is not divided evenly. Here are the areas of each district, given in square miles:

2941, 69491, 1021, 17181, 5362

Evaluate the median and the mean of this data set.

Solution

To evaluate the median, first sort the data into either increasing or decreasing order: 1021, 2941, 5362, 17181, 69491. Because there is an odd number of data values, we know that there must be one that is in the middle. The median area of the congressional districts is 5362 square miles.

To evaluate the mean, add the areas and then divide by 5, the number of districts:

$$\frac{2941 + 69491 + 1021 + 17181 + 5362}{5} = \frac{95996}{5}$$
$$= 19199.2 \text{ mi}^2$$

Notice that the mean of this data set is more than three times as much as the median. This happens in data sets where a small number of values are much greater than the rest. In this case, the second congressional district is about four times as large as the next biggest district.

There is a similar and even more striking effect when we compare the median city population with the *mean* city population among Oregon's 241 incorporated cities. Oregon's smallest incorporated city is Lonerock, with a population of 21 residents. The city with the median population is Cave Junction, at 1883 residents. There are exactly 121 cities with a population greater than or equal to 1883 and 121 cities with a population less than or equal to 1883.

While the median city population is 1883, the *mean* city population is 11,079. This is because there are many small cities and few big cities in Oregon. In fact, the nine biggest cities account for more than half of Oregon's incorporated population, and the 232 smallest cities account for less than half of Oregon's incorporated population.

When a data set is displayed in a histogram, we can identify the bin that contains the median value by adding up the bin heights from left to right to find the total number of data values. Keep track of the running total. You're looking for the bin in which the running total first exceeds half of the total number of data values. That bin will contain the median. Looking back at the histogram in Part A that presented the ages of Oregon residents, we estimate the median age of Oregon residents in Figure 2.

Bin	Bar Height	Running Total
younger than 5	223005	
between 5 and 14	476572	699577
between 15 and 24	474833	1174410
between 25 and 34	470695	1645105
between 35 and 44	526574	2171679
between 45 and 54	507155	2678834
between 55 and 64	304388	2983222
between 65 and 74	219342	3202564
between 75 and 84	161404	3363968
older than 84	57431	3421399

Figure 2

Half the total number of data values is $\frac{3,421,399}{2}$ = 1,710,699.5, and the running total first exceeds 1,710,699.5 in the "between 35 and 44" bin, so we know that the median age of Oregonians must be between 35 years and 44 years.

Practice B

Calculate or approximate the median of each data set. Turn the page and check your solutions.

7. {4, 5, 8, 7, 4}

8. {3, 3, 2, 6, 9, 7}

9. {24, 59, 31, 8, 125, 47, 92, 23}

10.

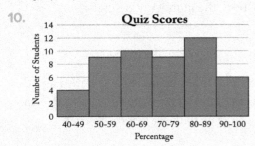

11.

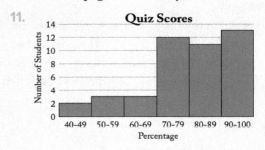

12.

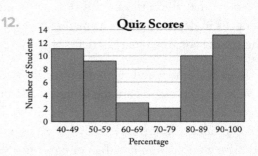

C. Mode is the Most Frequent

If any data value occurs more than once in a data set, then that data set has a **mode**. The mode is the data value that occurs most frequently within the data set — as long as the mode occurs at least twice. If there is a tie for most frequent data value, then each value in the tie is a mode. It's easier to spot repeated data values when they are displayed in either increasing or decreasing order, just as we displayed them to calculate the median. Here are the midterm exam scores from the beginning of this section written in decreasing order. The repeated values are highlighted:

96, 92, 88, 87, 86, 85, 84, 83, 82, 81, 79, 75,
73, 71, 71, 71, 69, 64, 64, 61, 60, 58, 58, 56

The most frequently occurring value is 71 with three occurrences, so it is the mode of this data set.

Example 3

Determine the mode(s) of the following data sets, if possible.

1. $\{3, 3, 4\}$

2. $\{10, 10, 15, 12, 15, 13, 18, 12, 15, 17, 12\}$

3. $\{5, 8, 16, 13, 22, 28, 33, 56\}$

Solution

1. The mode is 3.

2. The modes are 12 and 15.

3. There is no mode.

Practice C

Determine the mode(s) of the following data sets, if possible. Turn the page and check your solutions.

13. $\{4, 5, 8, 7, 4\}$ 16. $\{9, 2, 13, 9, 7, 2, 11, 8, 2\}$

14. $\{3, 3, 2, 6, 9, 7\}$ 17. $\{9, 2, 13, 9, 7, 2, 11, 8, 5\}$

15. $\{24, 59, 31, 8, 125, 47, 92, 23\}$ 18. $\{6, 3, 13, 9, 7, 2, 11, 8, 5\}$

Exercises 5.2

Complete parts A, B, and C for each of the following data sets. Give numbers to the tenths if they are not whole numbers.

A) Calculate the mean.

B) Calculate the median.

C) Determine the mode(s). If there is no mode, state this.

1. $1, 8, 8, 6, 5$ 4. $5, 2, 7, 9, 4$ 7. $3, 5, 8, 4$

2. $5, 9, 5, 7, 6$ 5. $7, 6, 2, 6, 2$ 8. $8, 7, 3, 8$

3. $4, 5, 9, 8, 7$ 6. $4, 7, 9, 7, 4$ 9. $6, 8, 3, 2$

Practice A — Answers

1. 5.6   4. 71.8

2. 5   5. 80

3. 51.125   6. 71.25

10. 8, 1, 7, 4

11. 8, 6, 3, 8

12. 3, 8, 6, 4

13. 8, 5, 7, 5, 2, 1

14. 7, 7, 2, 7, 9, 4

15. 9, 7, 9, 5, 7, 5, 2

16. 3, 6, 4, 9, 4, 2, 9, 4, 4, 4, 6

17. 2, 8, 3, 1, 9, 8, 8, 5

18. 3, 5, 1, 3, 6, 4, 9, 4, 2, 9, 4, 6, 1, 8

19. 3, 85, 65, 21, 95, 23, 2

20. 90, 75, 94, 96, 99

21. 964, 723, 368, 944, 837, 891, 492

22. 5792, 477, 8094, 1696, 971, 4754, 477, 3

23. The table below lists heights of players on the 2016 University of Oregon women's basketball team.

Player	Height (in.)
Lauren Yearwood	75
Morgan Yaeger	69
Jacinta Vandenberg	77
Megan Trinder	67
Mar'Shay Moore	68
Mallory McGwire	77
Sabrina Ionescu	70
Ruthy Hebard	76
Justine Hall	70
Lydia Giomi	78
Oti Gildon	73
Maite Cazorla	70
Sierra Campisano	75
Lexi Bando	69

24. The table below lists heights of players on the 2016 University of Oregon men's basketball team.

Player	Height (in.)
Jordan Bell	81
M.J. Cage	82
Keith Smith	79
Paul White	81
Dillon Brooks	79
Chris Boucher	82
Kavell Bigby-Williams	83
Roman Sorkin	82
Casey Benson	75
Payton Pritchard	74
Tyler Dorsey	76
Charlie Noebel	74
Evan Gross	70
Dylan Ennis	74

Practice B — Answers

7. 5

8. 4.5

9. 39

10. Between 70 and 79

11. Between 80 and 89

12. Between 70 and 79

25. The table below lists heights of players on the 2016 Oregon State University men's basketball team.

Player	Height (in.)
Cheikh N'diaye	84
Gligorije Rakocevic	83
Christian Russell	83
Tres Tinkle	80
Drew Eubanks	82
Matt Dahlen	78
Ben Koné	80
JaQuori McLaughlin	76
Stephen Thompson Jr.	76
Ronnie Stacy	76
Tanner Sanders	77
Daine Muller	76
Kendal Manuel	76

26. The below table lists the areas of each county in Connecticut.

County	Area (sq. mi.)
Fairfield	837
Hartford	750.6
Litchfield	944.6
Middlesex	439.1
New Haven	862.1
New London	771.7
Tolland	417
Windham	521.5

27. The densities of various materials are shown in grams per cubic centimeter.

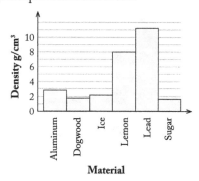

28. The hardnesses of five materials are shown using the Knoop Hardness Scale.

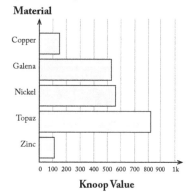

Practice C — Answers

13. The mode of $\{4, 5, 8, 7, 4\}$ is 4

14. The mode of $\{3, 3, 2, 6, 9, 7\}$ is 3

15. $\{24, 59, 31, 8, 125, 47, 92, 23\}$ has no mode.

16. The mode of $\{9, 2, 13, 9, 7, 2, 11, 8, 2\}$ is 2.

17. The modes of $\{9, 2, 13, 9, 7, 2, 11, 8, 5\}$ are 9 and 2.

18. $\{6, 3, 13, 9, 7, 2, 11, 8, 5\}$ has no mode.

29. The pictograph shows the frog populations for three lakes.

Lake Frog Population

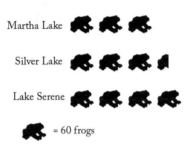

30. A player in the game Age of Empires II chooses a civilization to develop into an empire. The bar graph shows the number of players who chose to play each of five different civilizations in an online computer gaming tournament.

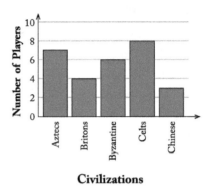

For the following exercises, complete parts A and B for each data set. Give numbers to the tenths if they are not whole numbers.
A) Approximate the mean.
B) Approximate the median.

31.

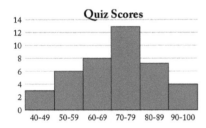

33.

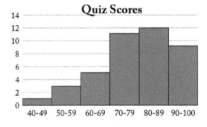

32.

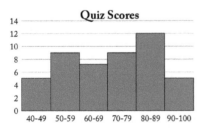

34.

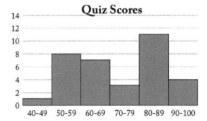

When calculating the mean, use the ages 2.5, 10, 20, 30, … , 90 to represent the age ranges.

35.

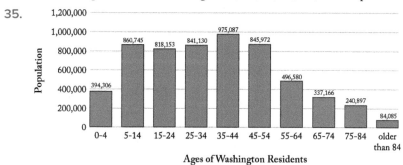

36.

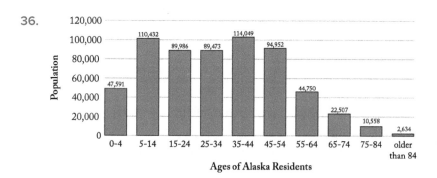

37.

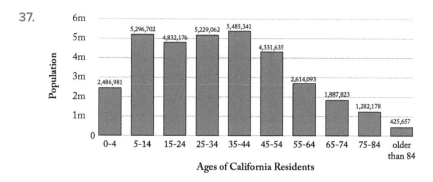

38.

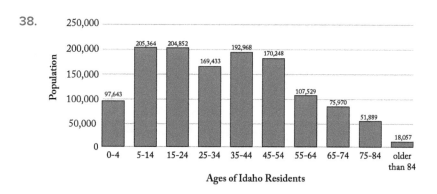

39. An instructor asks his students how long they spend brushing their teeth in the morning. When calculating the mean, use the brushing times 15, 45, 75, 105, 135 and 165 to represent time ranges.

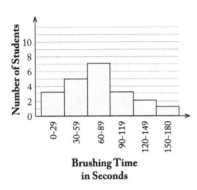

Brushing Time in Seconds

40. An instructor asks his students how many classes they are taking. The histogram shows the results.

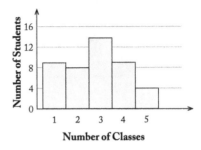

Number of Classes

For the following exercises, think about the exercise and then answer carefully.

41. Calculate the mean, median and mode for the data sets a) 3, 3, 2, 4, 3 and b) 3, 3, 2, 4, 3, 100. Give answers to the tenths if they aren't whole numbers. For the second data set we added an outlier. When the outlier was added, did the mean change? What about the median or mode?

42. A student says that 2 is the median of the data set 3, 3, 2, 4, 3, because it is in the middle, like the median of a freeway is right in the middle. "Look," the student says, "I can make 4 become the median by rearranging the values to 3, 3, 4, 2, 3." What would you say to this student?

43. Imagine a neighborhood with many cheap houses and one expensive mansion. If a real estate agent wants to give the impression that the neighborhood is full of nice and pricey houses, would they present the mean, median or mode to a prospective customer?

5.3 Equations with Two Variables

Xavier works in a restaurant for \$10 per hour, and his wife, Yvonne, works as a receptionist for \$12 per hour. One week they earned \$500 together. How many hours did Xavier work that week? How many hours did Yvonne work that week?

We can translate this scenario into an equation, but more than one possible pair of answers to its questions exist. Because there are two unknown quantities, we need two variables. If x is the number of hours that Xavier worked, then the expression $10x$ represents the amount of money that he earned. If y is the number of hours that Yvonne worked, then $12y$ represents the amount of money that she earned. Adding these expressions gives us one way to represent the amount of money that they earned together. We know that in this scenario, the amount of money that they earned together is the number 500, so this scenario can be translated into the equation $10x + 12y = 500$.

When an equation has more than one variable, a solution of the equation must include a value for each of the variables involved. In this section, we will look specifically at equations with two variables. For the sake of simplicity, we will only use the variables x and y in this section.

A solution of an equation may be given as an **ordered pair** of numbers.

Ordered Pair

An ordered pair is a set of two number values that are substituted for the variables in a specified order.

An ordered pair of numbers consists of two numbers separated by a comma and enclosed in parentheses, such as $(8, 3)$. A solution of an equation may also be given as two simple equations, such as $x = 8$ and $y = 3$, or as a row in a table:

x	y
8	3

A. Verifying Solutions

To verify that a given pair of values is a solution of an equation, substitute each variable with the given number value. The resulting equation consists of two arithmetic expressions separated by an equal sign. Evaluate the arithmetic expressions to determine whether the resulting equation is a true statement. If the equation is a true statement, then the pair of values is a solution of the equation. If the equation is a false statement, then the pair of values is not a solution of the equation.

Example 1

Determine whether each pair of values is a solution of the equation $5x - 7y = 13$

1. $x = 11$ and $y = 6$
2. $x = 4$ and $y = -1$
3. $x = -3$ and $y = -4$

Solutions

1. We substitute $x = 11$ and $y = 6$ into $5x - 7y = 13$ and determine whether the resulting equation is a true statement. Remember to follow the order of operations.

$$5 \cdot 11 - 7 \cdot 6 = 13$$
$$55 - 42 = 13$$
$$13 = 13$$

We end up with a *true* statement, so we have verified that $x = 11$ and $y = 6$ *is* a solution of $5x - 7y = 13$.

2. Substitute $x = 4$ and $y = -1$ into $5x - 7y = 13$ and determine whether the resulting equation is a true statement:

$$5 \cdot 4 - 7 \cdot (-1) = 13$$
$$20 - (-7) = 13$$
$$27 = 13$$

We end up with a *false* statement, so we have verified that $x = 4$ and $y = -1$ is *not* a solution of $5x - 7y = 13$.

3. Substitute $x = -3$ and $y = -4$ into $5x - 7y = 13$ and determine whether the resulting equation is a true statement:

$$5 \cdot (-3) - 7 \cdot (-4) = 13$$
$$-15 - (-28) = 13$$
$$13 = 13$$

We end up with a *true* statement, so we have verified that $x = -3$ and $y = -4$ *is* a solution of $5x - 7y = 13$.

Example 2

Determine whether each ordered pair is a solution of the equation $y = 3x - 8$. The ordered pairs each give the x variable as the first number and the y variable as the second number.

1. $(5, 7)$
2. $\left(\frac{17}{3}, 9\right)$
3. $\left(\frac{13}{4}, \frac{7}{4}\right)$

Solution

1. $(5, 7)$ *is* a solution of $y = 3x - 8$:

$$7 = 3 \cdot 5 - 8$$
$$7 = 15 - 8$$
$$7 = 7$$

2. $\left(\frac{17}{3}, 9\right)$ *is* a solution of $y = 3x - 8$:

$$9 = 3 \cdot \frac{17}{3} - 8$$
$$9 = 17 - 8$$
$$9 = 9$$

3. $\left(\frac{13}{4}, \frac{7}{4}\right)$ *is* a solution of $y = 3x - 8$:

$$\frac{7}{4} = 3 \cdot \frac{13}{4} - 8$$
$$\frac{7}{4} = \frac{39}{4} - 8$$
$$\frac{7}{4} = \frac{39}{4} - \frac{32}{4}$$
$$\frac{7}{4} = \frac{7}{4}$$

As the previous examples illustrate, equations with two variables have many solutions. A table provides a convenient way to write several solutions for an equation in one place. Values for one variable are listed in the left-hand column, with that variable as the column header. Values for the other variable are listed in the right-hand column, with that variable as the column header. Numbers in the same row are considered to be an ordered pair, which may or may not be a solution.

Example 3

Verify that each pair of values given in the table is a solution of the equation $x^2 - 2y = 2x + y$. Cross out any ordered pair that is not a solution.

x	y
-3	5
6	8
5	2
1	$-\frac{1}{3}$

Solution

$(-3, 5)$ *is* a solution:

$$(-3)^2 - 2 \cdot 5 = 2 \cdot (-3) + 5$$
$$9 - 10 = -6 + 5$$
$$-1 = -1$$

$(6, 8)$ *is* a solution:

$$6^2 - 2 \cdot 8 = 2 \cdot 6 + 8$$
$$36 - 16 = 12 + 8$$
$$20 = 20$$

$(5, 2)$ *is not* a solution:

$$5^2 - 2 \cdot 2 \neq 2 \cdot 5 + 2$$
$$25 - 4 \neq 10 + 4$$
$$21 \neq 14$$

Therefore, we cross out that pair on the table:

x	y
-3	5
6	8
~~5~~	~~2~~
1	$-\frac{1}{3}$

$\left(1, -\frac{1}{3}\right)$ *is* a solution:

$$1^2 - 2 \cdot \left(-\frac{1}{3}\right) = 2 \cdot 1 + \left(-\frac{1}{3}\right)$$
$$1 + \frac{2}{3} = 2 - \frac{1}{3}$$
$$\frac{5}{3} = \frac{5}{3}$$

Practice A

In Xavier and Yvonne's $500 work week, one of the following is a possible pair of answers to the questions posed at the beginning of the section and the others are not. Which is a possible pair of answers? When you are finished, turn the page to check your solutions.

1. Xavier worked 25 hours and Yvonne worked 20 hours.

2. Xavier worked 15 hours and Yvonne worked 30 hours.

3. Xavier worked 20 hours and Yvonne worked 25 hours.

4. Xavier worked 30 hours and Yvonne worked 20 hours.

B. Completing Solutions of Equations with Two Variables

When we are given a value for one of the variables in a two-variable equation, we may substitute that value into the equation. The resulting equation has just one variable. Any solution of the resulting one-variable equation completes a solution of the two-variable equation.

Example 4

Complete a solution of the equation $3x + 8y = 36$ given $x = 4$.

Solution

First, substitute $x = 4$ into $3x + 8y = 36$ to obtain the one-variable equation $3 \cdot 4 + 8y = 36$. Then solve this equation to complete the solution:

$$
\begin{aligned}
3 \cdot 4 + 8y &= 36 \\
12 + 8y &= 36 \\
8y &= 24 \\
y &= 3
\end{aligned}
$$

We find that $x = 4$ and $y = 3$ is a solution of the equation $3x + 8y = 36$.

Example 5

Complete the solutions of the equation $y = \frac{3}{4}x - 2$ that are given in the table.

x	y
-4	
	7
5	
	−3

Solution

Let's start with the first row. Because -4 is in the x column, we substitute $x = -4$ into $y = \frac{3}{4}x - 2$ and solve for y:

$$y = \frac{3}{4} \cdot (-4) - 2 \qquad \text{Substitute } x \text{ with } -4$$

$$y = -3 - 2 \qquad \text{Simplify}$$

$$y = -5 \qquad \text{Simplify}$$

We'll add that solution to our table and move on to the second row.

x	y
-4	-5
	7
5	
	-3

Because 7 is in the y column of the second row, we substitute $y = 7$ into $y = \frac{3}{4}x - 2$ and solve for x:

$$7 = \frac{3}{4}x - 2 \qquad \text{Substitute 7 for } y.$$

$$9 = \frac{3}{4}x \qquad \text{Add 2 to both sides.}$$

$$\frac{4}{3} \cdot 9 = \frac{3}{4}x \cdot \frac{4}{3} \qquad \text{Multiply both sides by the reciprocal } \frac{4}{3}.$$

$$12 = x$$

We add that solution to our table and move on to the third row.

x	y
-4	-5
12	7
5	
	-3

See if you can follow along the rest of the way.

$$y = \frac{3}{4} \cdot 5 - 2$$

$$y = \frac{15}{4} - 2$$

$$y = \frac{15}{4} - \frac{8}{4}$$

$$y = \frac{7}{4}$$

x	y
−4	−5
12	7
5	$\frac{7}{4}$
	−3

$$-3 = \frac{3}{4}x - 2$$
$$-1 = \frac{3}{4}x$$
$$-\frac{4}{3} = x$$

Here is the table with the completed solutions:

x	y
−4	−5
12	7
5	$\frac{7}{4}$
$-\frac{4}{3}$	−3

Practice A — Answers

1. $x = 25$ and $y = 20$ *is not* a solution of $10x + 12y = 500$, so the pair of answers is not possible.

2. $x = 15$ and $y = 30$ *is not* a solution of $10x + 12y = 500$, so the pair of answers is not possible.

3. $x = 20$ and $y = 25$ *is* a solution of $10x + 12y = 500$, so the pair of answers is possible.

4. $x = 30$ and $y = 20$ *is not* a solution of $10x + 12y = 500$, so the pair of answers is not possible.

Practice B

5. Complete the table of solutions for the equation $y = 2x - 4$.

x	y
-1	
0	
1	
2	

6. Complete the table of solutions for the equation $3x + y = 5$.

x	y
1	
2	
3	
4	

7. Complete the solutions for the equation $5x - 9y = 42$.

x	y
3	
	2
−4	
	−4

8. Remember Xavier and Yvonne? If Yvonne worked 20 hours, how many hours did Xavier work?

9. If Xavier worked 14 hours, How many hours did Yvonne work?

10. If Yvonne worked 15 hours, how many hours did Xavier work?

When you are finished, turn the page and check your solutions.

C. Creating Solutions of Equations with Two Variables

To create a solution of an equation with two variables, we need to choose a value for one variable and then complete the solution.

> ### Example 6
>
> Create a solution of the equation $4x + 5y = 61$.
>
> Solution
> There are an infinite number of correct solutions for this equation. Here are a few possibilities.
>
> 1. Edwardo began by choosing $x = 4$ and then completed the solution:
>
> $$4 \cdot 4 + 5y = 61$$
> $$16 + 5y = 61$$
> $$5y = 45$$
> $$y = 9$$
>
> Edwardo's solution is $x = 4$ and $y = 9$.

2. Jess began by choosing $y = -3$ and then completed the solution:

$$4x + 5 \cdot (-3) = 61$$
$$4x + (-15) = 61$$
$$4x = 76$$
$$x = 19$$

Jess's solution is $x = 19$ and $y = -3$.

3. Lydia began by choosing $y = 4$ and then completed the solution:

$$4x + 5 \cdot 4 = 61$$
$$4x + 20 = 61$$
$$4x = 41$$
$$x = \frac{41}{4}$$

Lydia's solution is $x = \frac{41}{4}$ and $y = 4$.

Practice C

11. Create your own solution to the equation $y = 2x + 4$. To check your answer, substitute the values for x and y that you end up with into the equation and verify that you end up with a true statement.

12. Create your own solution to the equations $4x + 5y = 61$. Check your answer as you did with question 11.

13. Returning to Xavier and Yvonne, create your own solution to the equation $10x + 12y = 500$ to find one possible pair of answers to the questions "How many hours did Xavier work that week? How many hours did Yvonne work that week?" Since it doesn't make sense for either Xavier or Yvonne to work a negative number of hours, make sure that neither value in your solution is negative. When you are finished, verify your solutions.

Exercises 5.3

For the following exercises, determine whether each pair of values is a solution of the equation.

1. $2x + 3y = 11$
$x = 4$ and $y = 1$

2. $4x + 3y = 21$
$x = 3$ and $y = 3$

3. $y = 3x - 4$
$x = 3$ and $y = 6$

4. $y = \frac{2}{3}x - 5$
$x = 6$ and $y = 9$

5. $2x - 3y = 28$
$x = 5$ and $y = 6$

6. $5x - 4y = 39$
$x = 3$ and $y = 6$

7. $x - 2y = 8$
   $x = 4$ and $y = -2$

8. $3x - y = 12$
   $x = 3$ and $y = 5$

For the following exercises, determine whether each ordered pair is a solution of the equation.

9. $5x - 2y = 14$
   $(4, 3)$

11. $4x - y = 12$
    $\left(\frac{23}{6}, \frac{10}{3}\right)$

13. $y = \frac{4}{3}x + 2$
    $\left(4, \frac{3}{2}\right)$

10. $6x - 2y = 18$
    $(4, 3)$

12. $x - 3y = 8$
    $\left(\frac{7}{2}, -\frac{3}{2}\right)$

14. $y = \frac{2}{3}x + \frac{3}{4}$
    $\left(\frac{7}{8}, \frac{3}{2}\right)$

Practice B — Answer

5.

x	y
-1	-6
0	-4
1	-2
2	0

6.

x	y
1	2
2	-1
3	-4
4	-7

7.

x	y
3	-3
12	2
-4	$\frac{62}{9}$
$\frac{6}{5}$	-4

8. Complete a solution of the equation $10x + 12y = 500$ given $y = 20$:

$$10x + 12 \cdot 20 = 500$$
$$10x + 240 = 500$$
$$10x = 260$$
$$x = 26$$

Xavier worked 26 hours.

9. Complete a solution of the equation $10x + 12y = 500$ given $x = 14$:

$$10 \cdot 14 + 12y = 500$$
$$140 + 12y = 500$$
$$12y = 360$$
$$y = 30$$

Yvonne worked 30 hours.

10. Complete a solution of the equation $10x + 12y = 500$ given $y = 15$:

$$10x + 12 \cdot 15 = 500$$
$$10x + 180 = 500$$
$$10x = 320$$
$$x = 32$$

Xavier worked 32 hours.

15. $y = -\frac{2}{3}x + 2$

    $(3, 4)$

16. $y = -\frac{3}{4}x + \frac{7}{2}$

    $(-2, 2)$

For each of the following exercises, complete the table of solutions of the equation.

17. $y = 3x + 2$

x	y
0	
	5
4	
	20

20. $4x + 2y = 3$

x	y
$-\frac{3}{4}$	
	$\frac{3}{2}$
	$\frac{1}{2}$
$\frac{3}{4}$	

23. $y = -2x + 1$

x	y
-2	
0	
	0
2	

18. $y = 2x - 5$

x	y
-2	
	1
3	
	11

21. $y = 3x$

x	y
0	
$\frac{1}{3}$	
$\frac{2}{3}$	
$\frac{3}{4}$	

24. $y = 2x - 1$

x	y
-2	
0	
	0
2	

19. $x + y = -5$

x	y
	-2
-1	
0	
	-7

22. $y = 1\_3\,x$

x	y
0	
	$\frac{1}{3}$
6	
	$\frac{7}{6}$

For each of the following exercises, create your own solution to the equation. To check your answer, substitute the values for x and y that you end up with into the equation and verify that you end up with a true statement.

25. $y = 3x - 1$

26. $y = x - 3$

27. $y = \frac{4}{3}x$

28. $y = 5x$

29. $4x - 2y = 15$

30. $-5x + 2y = 14$

31. $y = -\frac{3}{5}x + \frac{4}{5}$

32. $y = \frac{3}{4}x + \frac{5}{2}$

Practice C — Answer

Answers may vary.

For each of the following exercises, translate the scenario into an equation.

33. Xander and Yasmin find $12 under the cushions of their couch. They decide to split the money, not necessarily equally. Let x represent the amount of money that Xander keeps, and let y represent the amount of money that Yasmin keeps.

34. At a tropical fish store, the X-Ray Tetra cost $2 each, and the Yellow Tail Acei Cichlid cost $9 each. Let x represent the amount of money that a customer spends on X-Ray Tetras, and let y represent the amount of money that he spends on Yellow Tail Acei. The customer spends $41 total.

35. Ben sells his old toys at a garage sale for $47 total. He sells his Xbox games for $5 each, and his yoyos for $3 each. Let x represent the number of Xbox games sold, and let y represent the number of yoyos sold.

36. Mary is a computer programmer. She writes code using the Xquery and YQL languages. She recently wrote 1200 lines of code. Let x represent the number of lines of Xquery code, and let y represent the number of lines of YQL code she wrote.

5.4 The Cartesian Plane

In earlier chapters, we worked with the number line as a geometric model for the set of real numbers. **Real numbers** are a set of numbers that includes every rational number, but also many other numbers that are not rational. Every real number is represented uniquely by one point on the number line, and every point on the number line uniquely represents one real number.

In the previous section, we learned that solutions of equations with two variables are called ordered pairs of numbers. There is a geometric model similar to the number line for ordered pairs called the **Cartesian plane**. The Cartesian plane is named in honor of Rene Descartes, an influential French scientist, philosopher, and mathematician. Though it might appear tricky at first, the **Cartesian plane** makes solving some problems much easier.

A. Plotting Ordered Pairs on the Cartesian Plane

The **Cartesian plane** is defined by two number lines that are perpendicular to each other and cross at their zero points. These number lines are called the **axes** (the plural of *axis*, pronounced "ax-EEZ") of the Cartesian plane. Usually, one axis is oriented horizontally with numbers increasing as they go to the right, and the other axis is oriented vertically with numbers increasing as they go up. The point where the axes cross is called the **origin**. The axes divide the plane into four regions, called **quadrants**.

The region above the horizontal axis and to the right of the vertical axis is the **first quadrant**, or QI for short. The region above the horizontal axis and to the left of the vertical axis is the **second quadrant**, or QII for short. The region below the horizontal axis and to the left of the vertical axis is the **third quadrant**, or QIII for short. The region below the horizontal axis and to the right of the vertical axis is the **fourth quadrant**, or QIV for short (Figure 1).

Every ordered pair of real numbers is represented uniquely by one point on the Cartesian plane, and every point on the Cartesian plane uniquely represents one ordered pair of real numbers.

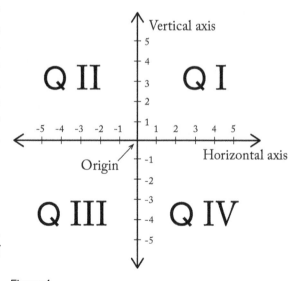

Figure 1

If a and b are any two real numbers, then the point that represents the ordered pair (a, b) is directly above or below the number a on the horizontal axis and directly across from the number b on the vertical axis.

Every ordered pair that has a for the first number is represented by a point on the *vertical* line that goes through the number a on the horizontal axis. Also, every ordered pair that has b for the second number is represented by a point on the *horizontal* line that goes through the number b on the vertical axis. That means that the only possible location for the point that represents the ordered pair (a, b) is the point where those two lines cross (Figure 2).

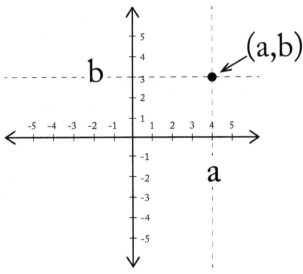

Figure 2

It's important to remember that the first number in an ordered pair represents a location on the horizontal axis, and the second number represents a location on the vertical axis. To help you remember, we'll remind you that the words "horizontal" and "vertical" appear in alphabetical order. The horizontal comes first and the vertical second.

Example 1

Plot and label the following ordered pairs on the Cartesian plane. Then identify the quadrant in which the point is located.

1. $(6, \frac{5}{3})$
2. $(0, 4)$
3. $(-5, -1)$
4. $(6, 0)$
5. $(4, -\frac{5}{2})$
6. $(-4, 3)$

Solution

1. $(6, \frac{5}{3})$ is in QI.

2. $(0, 4)$ is on the vertical axis.

3. $(-5, -1)$ is in QIII.

4. $(6, 0)$ is on the horizontal axis.

5. $(4, -\frac{5}{2})$ is in QIV.

6. $(-4, 3)$ is in QII.

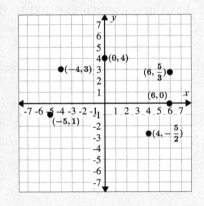

Practice A

Plot and label the following ordered pairs on the Cartesian plane. Then identify the quadrant in which the point is located. When you're finished, turn the page to check your answers.

1. $(2, 3)$
2. $(-2, 4)$
3. $(0, 0)$
4. $(2, -\frac{3}{2})$
5. $(-\frac{8}{5}, -\frac{4}{3})$
6. $(0, -1)$

B. Plotting Solutions of Two-Variable Equations

When we dealt with the number line, you learned that the arrows on both ends indicate that the line carries on infinitely in both directions. The same is true for the Cartesian plane and its horizontal and vertical axes. We have to limit what is visible of the Cartesian plane because we can't show an infinite plane. The part of the plane that you can see in a plot is known as the **window**.

When an ordered pair represents a solution of an equation with two variables, the variable for the first number in the ordered pair is associated with the horizontal axis. We label the horizontal axis with this variable near the arrowhead at the right-hand side of the window. The variable for the second number in the ordered pair is associated with the vertical axis. We label the vertical axis with this variable near the arrowhead at the top of the window.

The numbers in the ordered pair are called **coordinates**, and each coordinate is named by the associated variable. For example, the point in Figure 3 labeled B represents ordered pair $(1, -2)$. As always, the first number represents a position on the horizontal axis. Because it is labeled with the variable x, we call the horizontal axis the **x-axis**, and we call the first number in the ordered pair the x-coordinate of the point B. In the same way, the vertical axis is labeled with the variable y, so it is called the **y-axis** and the second number of any ordered pair that is plotted on this plane is called the y-coordinate of the point. So the x-coordinate of point B is 1 and the y-coordinate of point B is -2.

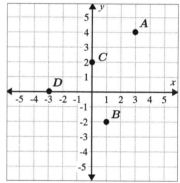

Figure 3

Example 2

Use Figure 3 to answer the following questions.

1. What is the x-coordinate of point D?

2. What is the y-coordinate of point A?

3. What ordered pair is represented by the point C?

4. Each of these points represents an ordered pair that may or may not be a solution of the equation $2x - 3y = -6$. Which points are solutions? Which points are not solutions?

Solutions

1. The x-coordinate of point D is –3.

2. The y-coordinate of point A is 4.

3. The point C represents the ordered pair $(0, 2)$.

4. Point A represents the ordered pair $(3, 4)$. Substituting $x = 3$ and $y = 4$ into $2x - 3y = -6$, we get a true statement: $2 \cdot 3 - 3 \cdot 4 = -6$. Point A *is* a solution.

   Point B represents the ordered pair $(1, -2)$. Substituting $x = 1$ and $y = -2$ into $2x - 3y = -6$, we get a false statement: $2 \cdot 1 - 3 \cdot (-2) = -6$. Point B is not a solution because the left and right sides are not equal.

   Point C represents the ordered pair $(0, 2)$. Substituting $x = 0$ and $y = 2$ into $2x - 3y = -6$, we get a true statement: $2 \cdot 0 - 3 \cdot 2 = -6$. Point C is a solution.

   Point D represents the ordered pair $(-3, 0)$. Substituting $x = -3$ and $y = 0$ into $2x - 3y = -6$, we get a true statement: $2 \cdot (-3) - 3 \cdot 0 = -6$. Point D is a solution.

Practice B

Use the graph to answer the following questions. When you're finished, turn the page to check your answers.

7. What is the x-coordinate of point B?

8. What is the y-coordinate of point C?

9. What ordered pair is represented by the point A?

10. Which point has the coordinates $(-2, -1)$?

11. Each of these points represents an ordered pair that may or may not be a solution of the equation $y = \frac{1}{2}x$? Which points are solutions? Which points are not solutions?

12. Each of these points represents an ordered pair that may or may not be a solution of the equation $x + 4y = 14$. Which points are solutions? Which points are not solutions?

Practice A – Answers

1. $(2, 3)$ is in QI.

2. $(-2, 4)$ is in QII.

3. $(0, 0)$ is on the origin.

4. $(2, -\frac{3}{2})$ is in QIV

5. $(-\frac{8}{5}, -\frac{4}{3})$ is in QIII

6. $(0, -1)$ is on the vertical axis.

Exercises 5.4

For the following exercises, identify the quadrant in which each point is located. However, if a point is on an axis, then identify the axis as vertical or horizontal. Alternately, if a point is at the origin, indicate this.

1.

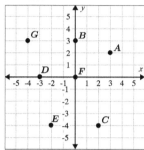

2.

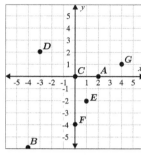

For the following exercises, give the ordered pair corresponding to each point.

3.

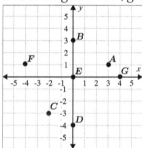

4.

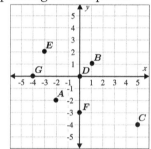

For the following exercises, plot the points on a Cartesian plane like the one below, and label each with the corresponding capital letter.

5.   Point A, $(-4, 3)$; Point B, $(0, 2)$;
     Point C, $(-3, 0)$; Point D, $(3, -4)$;
     Point E, $(0, 0)$; Point F, $\left(4, \frac{5}{2}\right)$

6.   Point A, $(2, 3)$; Point B, $(-3, -3)$;
     Point C, $(1, 1)$; Point D, $(-4, 0)$;
     Point E, $(0, -4)$; Point F, $\left(-\frac{3}{2}, \frac{1}{2}\right)$

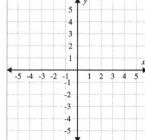

Answer the following questions. For these exercises, the horizontal axis is the x-axis, and the vertical axis is the y-axis.

7.   What is the x-coordinate of the ordered pair $\left(-\frac{3}{2}, \frac{1}{2}\right)$?

8.   What is the y-coordinate of the ordered pair $(3, -4)$?

9.   What is the y-coordinate of the ordered pair $(2, 7)$?

10. What is the x-coordinate of the ordered pair $(7.8, 4.5)$?

11. What is the x-coordinate of the ordered pair $(0, 3)$?

12. What is the y-coordinate of the ordered pair $(-2, -3)$?

13. What is the y-coordinate of the ordered pair $(4, 0)$?

14. What is the x-coordinate of the ordered pair $\left(3, \frac{3}{4}\right)$?

15. In the graph below, which point does *not* satisfy the linear equation $y = 2x - 1$? Substitute its coordinates into the equation to verify that the result is not a true statement.

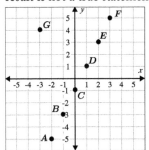

16. In the graph below, which point does *not* satisfy the linear equation $y = \frac{1}{2}x - 1$? Substitute its coordinates into the equation to verify that the result is not a true statement.

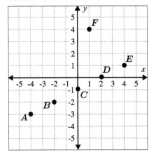

17. In the graph below, which point does *not* satisfy the linear equation $-2x - y = 4$? Substitute its coordinates into the equation to verify that the result is not a true statement.

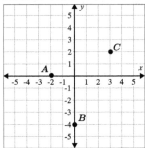

18. In the graph below, which point does *not* satisfy the linear equation $2x - 3y = 4$? Substitute its coordinates into the equation to verify that the result is not a true statement.

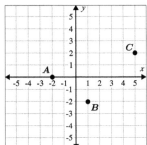

19. When a point does not lie on an intersection of a grid, it is common to estimate a coordinate to one tenth of a division between marks by eye. We imagine the side of a square being divided into ten equal pieces. For example, Point A on the graph below could be estimated using the human eye to have the coordinates $(-3.0, -2.6)$. Estimate the coordinates for the other points to the tenths, giving your answers in ordered pair notation and identifying them using the corresponding capital letters.

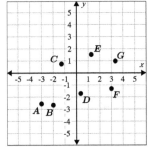

20. On a reasonably sized grid, we can plot points with coordinates given to the tenths of a division between marks. We imagine the side of a square being divided into ten equal pieces. For example, the number 2.6 is given to the tenths, so it can be used nicely as a coordinate on the Cartesian coordinate system shown, since the grid is large and the division between marks is one unit. Examine Point A, $(2.6, 2.3)$. Plot all of the points on a Cartesian plane (such as the one used for exercises 5 and 6). Label them with the corresponding capital letters.

    Point A. $(2.6, 2.3)$

    Point B. $(-3.8, 3.2)$

    Point C. $(-3.1, 0.0)$

    Point D. $(3.5, -3.5)$

    Point E. $(0.9, 0.9)$

    Point F. $(0.7, -3.4)$

21. One ordered pair in exercise 5 and one in exercise 6 had a coordinate that was an improper fraction. Before plotting a point, it is often helpful to change improper fractions to decimals rounded to the tenth of a division between tick marks. Plot the following points on a Cartesian plane (such as the one used for exercises 5 and 6), and label them with the corresponding capital letters. To do this, first change the improper fractions to decimals to the tenths.

    Point A. $\left(\frac{10}{3}, \frac{7}{4}\right)$

    Point B. $\left(\frac{13}{5}, -\frac{9}{2}\right)$

    Point C. $\left(-\frac{17}{7}, -\frac{27}{8}\right)$

    Point D. $\left(-\frac{25}{6}, \frac{12}{5}\right)$

22. Is it easier to determine the coordinates of Point A in Graph 1 or Graph 2? Why? A wise professor once said in a math department staff meeting, "Good communication isn't just making it technically possible for the reader to figure something out." Therefore, in the future you should label your points if it helps the reader, and if you have space to do it.

Graph 1

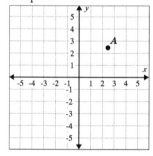

Graph 2

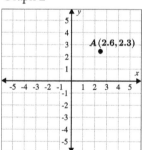

Practice B – Answers

7. The x-coordinate of point B is -2.

8. The y-coordinate of point C is 3.

9. The ordered pair $(-2, 4)$ is represented by the point A.

10. Point B has the coordinates $(-2, -1)$.

11. Points B and D are solutions to the equation $y = \frac{1}{2}x$. Points A and C are not.

12. Points A and C are solutions to the equation $x + 4y = 14$. Points B and D are not.

23. Which of the following scales are reasonable, and which are unreasonable? Why or why not?

a.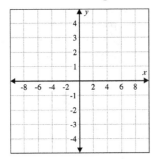

c.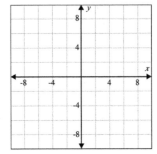

b.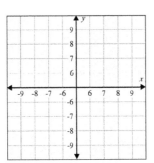

d.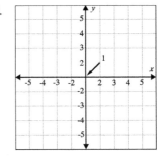

e.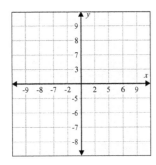

5.5 Graphs of Linear Equations

The table below displays several solutions of the equation $-3x + 5y = 12$.

x	y
-4	0
$-\frac{2}{3}$	2
1	3
4	4.8

These ordered pairs are plotted in this graph:

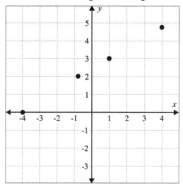

Do you notice anything interesting about these points? Look closely and you'll see that they appear to line up with each other. The technical name for this relationship is **collinearity**. We can also say that these points are **collinear** because there is a single straight line that goes through all of the points. In fact, *all* of the solutions of this equation — not just these four — plot to points that lie on this line. This equation belongs to a special class of equations called **linear equations**.

A. Linear Equations

Some two-variable equations have the same property that we observed above in the equation $-3x + 5y = 12$: when plotted on the Cartesian plane, the solutions are collinear. Here is the formal definition for this class of equations:

Linear Equations

Let a, b and c be any three constants such that a and b are not both equal to zero. Then the equation $ax + by = c$ is a **linear equation** with variables x and y. In addition, any equation that is equivalent to this equation is also a linear equation with variable x and y.

Example 1

Which of the following equations are linear equations?

1. $3x - 2y = 7$

2. $x = 8$

3. $2(3x - 5) = \dfrac{8 - 4y}{7}$

4. $3x^2 + 2y^2 = 7$

5. $2l + 2w = 40$

6. $2x + 5y = 3xy$

7. $y = \dfrac{3}{5}x + \dfrac{12}{5}$

8. $y = 7$

Solutions

1. $3x - 2y = 7$ *is* a linear equation. It is equivalent to $3x + (-2)y = 7$

2. $x = 8$ *is* a linear equation. It is equivalent to $1x + 0y = 8$

3. $2(3x - 5) = \dfrac{8 - 4y}{7}$ *is* a linear equation. It is equivalent to $42x + 4y = 78$

4. $3x^2 + 2y^2 = 7$ *is not* a linear equation because it is not equivalent to $ax + by = c$!

5. $2l + 2w = 40$ *is* a linear equation in the variables l and w.

6. $2x + 5y = 3xy$ *is not* a linear equation because it is not equivalent to $ax + by = c$!

7. $y = \dfrac{3}{5}x + \dfrac{12}{5}$ *is* a linear equation. It is equivalent to $-3x + 5y = 12$

8. $y = 7$ *is* a linear equation. It is equivalent to $0x + 1y = 7$

Not all of these solutions are easy to see like numbers 1 and 5. Let's take a minute to verify the solutions for the tough ones like numbers 3 and 7.

$$2(3x - 5) = \frac{8 - 4y}{7} \qquad \text{This is the original equation.}$$

$$7 \cdot 2(3x - 5) = \frac{8 - 4y}{7} \cdot 7 \qquad \text{Apply the multiplication principle.}$$

$$14(3x - 5) = 8 - 4y \qquad \text{Apply the distributive property.}$$

$$42x - 70 = 8 - 4y$$

$$42x - 70 + 70 = 8 - 4y + 70$$ Apply the addition principle.

$$42x = 78 - 4y$$

$$42x + 4y = 78 - 4y + 4y$$

$$42x + 4y = 78$$

Next, we'll check problem 7:

$$y = \frac{3}{5}x + \frac{12}{5}$$

$$5 \cdot y = 5 \cdot \left(\frac{3}{5}x + \frac{12}{5}\right)$$ Apply the multiplication principle and the distributive property.

$$5y = 5 \cdot \frac{3}{5}x + 5 \cdot \frac{12}{5}$$

$$5y = 3x + 12$$

$$-3x + 5y = 12$$ Apply the addition principle.

Practice A

Which of the following equations are linear equations? Turn the page to check your answers.

1. $2x + 3y = 4$

2. $y = -\frac{2}{3}x + \frac{5}{3}$

3. $y = x^2$

4. $2\left(x - \frac{3}{2}y\right) = 4$

5. $\frac{1}{xy} = 4$

6. $x + y = 7 - x - 8y$

B. The Graph of an Equation

If we were to plot every solution of an equation on the Cartesian plane the resulting collection of points is called the **graph of the equation**. The graph of a linear equation will always be a straight line. Since all of the solutions of a linear equation are collinear, we know that there is a straight line that goes through all of the solutions. It turns out that *every point* on that line represents an ordered

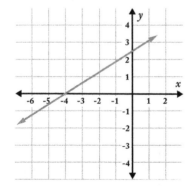

Figure 1.

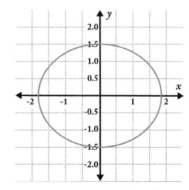

Figure 2. Graph of $3x^2 + 2y^2 = 7$

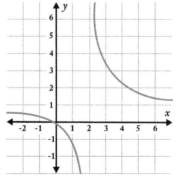

Figure 3. Graph of $2x + 5y = 3xy$

pair that is a solution of the equation. For example, the line in Figure 1 is the graph of the equation $-3x + 5y = 12$.

Notice that each of the ordered pairs listed in the table at the beginning of this section plot to points on this line. It's impossible to show the entire line because it extends infinitely far in both directions, but we indicate that the line segment we are showing is intended to represent the entire, infinite line by putting arrowheads at both ends of the line segment.

In Figures 2 and 3, we see the graphs of the equations in numbers 4 and 6 from the previous example. These equations were *not* linear, and we can clearly see that these graphs are not straight lines.

Practice B

7. Is the following graph linear?

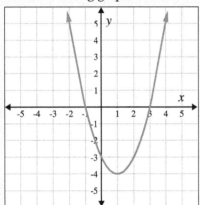

8. Is the following graph linear?

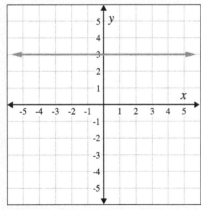

9. Is the following graph linear?

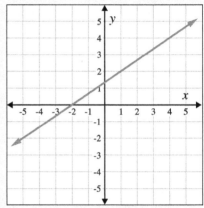

10. Does the point (2, 3) lie on the diagonal line shown in the graph above?

11. How about the point (-5, -2)?

12. Is the following an adequate graph of a linear equation?

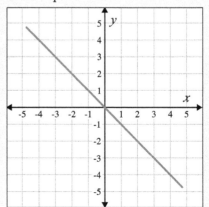

C. Graphing Linear Equations

When we're talking about equations, we can use the word "graph" as either a thing or an action:

- ◆ As a thing: "This line is the graph of the equation."

- ◆ As an action: "When we graph the equation, we draw this line."

To **graph** an equation is to construct its graph. Because the graph of a two-variable equation is a visual representation of all of its solutions, graphing a two-variable equation is really solving that equation. Because we know that the graph of a linear equation is a straight line, we can construct the graph by plotting two solutions and then using a ruler to draw the line that goes through those two points. To guard against the possibility of making a mistake, we will plot three solutions. The following example illustrates why this is a good idea.

▶ **Example 2**

Jose and Tran are both trying to graph linear equations. They each begin by creating three solutions of their equation.

1. Jose's solutions:

x	y
3	5
1	6
5	4

2. Tran's solutions:

x	y
2	5
1	7
3	4

Plot each of the points and then draw the straight line that goes through all three points to complete the graphs of the linear equations.

Practice A — Answers

1. $2x + 3y = 4$ is a linear equation. It is already in the form $ax + by = c$.

2. $y = -\frac{2}{3}x + \frac{5}{3}$ is a linear equation. It is equivalent to $2x + 3y = 5$, which is in the form $ax + by = c$.

3. $y = x^2$ is not a linear equation because is not equivalent to $ax + by = c$.

4. $2(x - \frac{3}{2}y) = 4$ is a linear equation. It is equivalent to $2x + (-3)y = 4$, which is in the form $ax + by = c$.

5. $\frac{1}{xy} = 4$ is not a linear equation because it is not equivalent to $ax + by = c$.

6. $x + y = 7 - x - 8y$ is a linear operation. It is equivalent to $2x + 9y = 7$, which is in the form $ax + by = c$.

Solutions

3. These three points are collinear. Here is the graph of the equation.

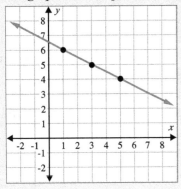

4. These three points are *not* collinear. Because the equation is linear, Tran must have made a mistake when he created his solutions. He needs to go back and look at his work again.

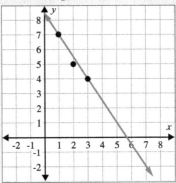

In Example 2, Tran would never have realized that he made a mistake if he had only created two solutions Because there will always be a line that goes through any two given points.

Here is the process that we follow to graph a linear equation:

1. Create *three* solutions of the equation.

2. Plot those three points.

3. If the three points *are* collinear, draw the straight line that goes through all three points. Congratulations! You are done. That line is the graph of the equation.

4. If the three points are *not* collinear, review your work in the first two steps to find the error. After fixing the error, proceed with step three.

Practice B — Answers

7. No, it is not linear. It is not a straight line.

8. Yes, it is linear. It is a straight line with arrowheads.

9. Yes, it is linear. It is a straight line with arrowheads.

10. The point (2, 3) does not lie on the line.

11. The point (-5, -2) does lie on the line.

12. No, it is not adequate. Although it is straight, it does not have arrowheads on the ends.

Example 3

Graph the following linear equations.

1. $x - y = 4$
2. $y = \frac{2}{3}x + 4$

Solutions

1. Here are three solutions of $x - y = 4$.

x	y
0	−4
4	0
2	−2

The shaded numbers are starting values that we chose, and the unshaded numbers in each row are the numbers we calculated to complete the solutions. When we plot these points, we see that they are collinear, so we draw the line that goes through the points.

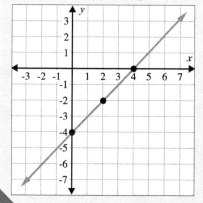

2. Here are three solutions of $y = \frac{2}{3}x + 4$.

x	y
0	4
−6	0
−3	2

The shaded numbers are starting values that we chose, and the unshaded numbers are the numbers we calculated to complete the solutions. When we plot these points, we see that they are collinear, so we draw the line that goes through the points.

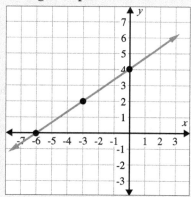

Notice that in both examples we created the first solution by choosing zero for the x-coordinate. The resulting point must be on the y-axis. That point is called the **y-intercept** of the line, the point where the line crosses the y-axis. Similarly, we created the second solution by choosing zero for the y-coordinate. The resulting point must be on the x-axis. That point is called the **x-intercept** of the line, the point where the line crosses the x-axis.

Practice C

13. Complete the table of solutions for the equation. then graph the equation.

$y = 2x - 4$

x	y
0	
1	
2	

14. Complete the table of solution for the equation. Then graph the equation.

$2x + 3y = -3$

x	y
-3	
0	
3	

15. Complete the table of solution for the equation. Then graph the equation. State the coordinates of the x-intercept and the y-intercept.

$2x - 4y = 8$

x	y
0	
	-1
	0

Graph the following equations. State the coordinates of the x-intercept and the y-intercept.

16. $y = 2x - 2$ **17.** $y = -\frac{5}{4}x + 3$ **18.** $x + 2y = 8$

When you are finished, turn the page to check your answers.

D. Horizontal and Vertical Lines

Near the beginning of this section, we defined linear equation and then followed up with an example consisting of eight equations that either were or were not linear equations. Two of those eight that were linear equations were $x = 8$ and $y = 7$. While these equations are linear, as demonstrated in that example, they are different in one important way from the other examples of linear equations that we've seen up to this point — they aren't *two*-variable equations.

The equation $x = 8$ places no restrictions on the y-coordinate of any solution, but the x-coordinate of every solution must be 8. Otherwise the equation $x = 8$ becomes a false statement. So the graph of this equation consists of every point of the Cartesian plane that has 8 as its x-coordinate, as you can see in Figure 4.

This line is a **vertical line**, parallel to the y-axis. Notice that the x-intercept of this line is the

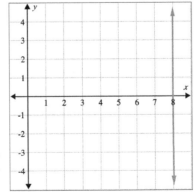

Figure 4.

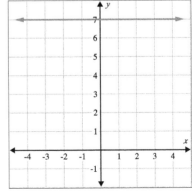

Figure 5.

point $(8, 0)$, but it has no y-intercept because it has no solution with an x-coordinate of zero.

In the same way, the equation $y = 7$ places no restrictions on the x-coordinate of any solution, but the y-coordinate of every solution must be 7. Otherwise, the equation $y = 7$ becomes a false statement. The graph of this equation consists of every point of the Cartesian plane that has 7 as its y-coordinate, as illustrated in Figure 5.

This line is a **horizontal line**, parallel to the x-axis. Notice that the y-intercept of this line is the point $(0, 7)$, but it has no x-intercept because it has no solution with a y-coordinate of zero.

Horizontal and Vertical Lines

Let k be any constant. The graph of the equation $x = k$ is a **vertical line**, and the graph of the equation $y = k$ is a **horizontal line**.

Example 4

Identify each equation as a vertical or horizontal line, or as neither vertical nor horizontal.

1. $x = -5$
2. $y - 3 = 0$
3. $x + y = 5 + y$
4. $y = 0$
5. $y = x$
6. $2x + 3 = 9$

Solutions

1. $x = -5$ is a vertical line.
2. $y - 3 = 0$ is the horizontal line $y = 3$.
3. $x + y = 5+y$ is the vertical line $x = 5$.
4. $y = 0$ is a horizontal line, in fact, $y = 0$ is the x-axis.
5. $y = x$ is neither horizontal nor vertical.
6. When we solve $2x + 3 = 9$ we end up with the equivalent equation $x = 3$, which is a vertical line.

Practice D

Answer the following questions about the equation $y = 3$.

19. Is the graph of the equation a vertical line, horizontal line, or neither vertical nor horizontal?

20. Graph the equation.

21. Give the x- and/or y-intercepts if there are any.

Answer the following questons about the equation $x = -4$.

22. Is the graph of the equation a vertical line, horizontal line, or neither vertical nor horizontal?

23. Graph the equation.

24. Give the x- and/or y-intercepts if there are any.

When you're finished, turn the page to check your work.

Practice C — Answers

13.

x	y
0	-4
1	-2
2	0

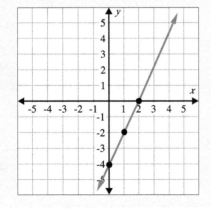

14.

x	y
-3	1
0	-1
3	-3

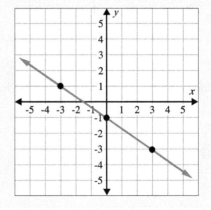

15.

x	y
0	-2
2	-1
4	0

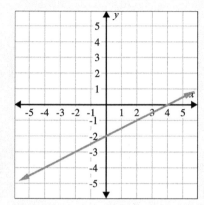

16.

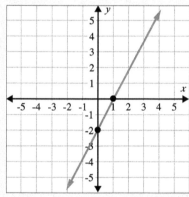

17. The x-intercept is $(\frac{12}{5}, 0)$, and the y-intercept is $(0, 3)$.

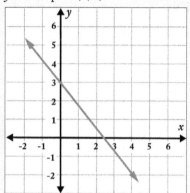

18. The x-intercept is $(8, 0)$ and the y-intercept is $(0, 4)$.

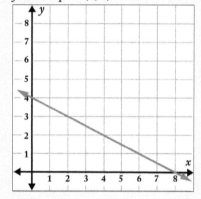

Exercises 5.5

For the following exercises, identify the graphs as linear or not linear.

1.

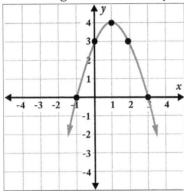

2.

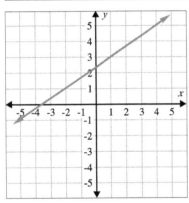

3.

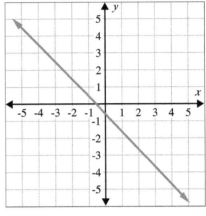

4.

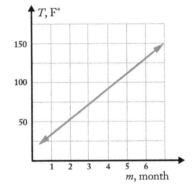

Practice D — Answers

19. $y = 3$ is a horizontal line.

20.

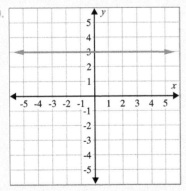

21. There is no x-intercept, but the y-intercept is $(0, 3)$.

22. $x = -4$ is a vertical line.

23.

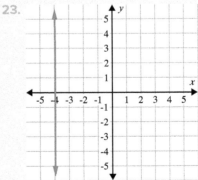

24. The x-intercept is $(-4, 0)$, but there is no y-intercept.

5.

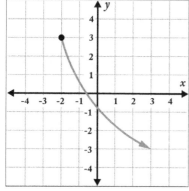

6.

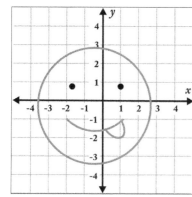

7.

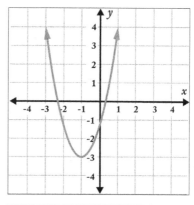

8.

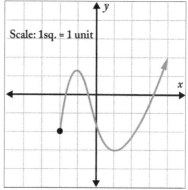

For the following exercises, verify that the equation is linear by putting it in the form $ax + by = c$, where a and b are not both zero.

9. $2y + 4 = x - 4$

10. $3(x - 1) + 2y = 7 - x$

11. $x - 3 + 2y = 0$

12. $5x - 2y - 3 = 2x + 4$

13. $\frac{2}{3}x + \frac{y - 6}{4} = 2 - \frac{y}{2}$

14. $3x + 2y + 4 = 2(x + y + 2)$

15. $3(x + 3y) - 2y = 4y + 5$

16. $2(2x + 3y) = 2(y - x) + 8$

For the following exercises, graph each linear equation by first completing the table to find three solutions to the equation, then plot the three points, and then finish graphing the line. Don't forget to use arrowheads, to label your axes, and to show your scale.

17. $y = 2x - 1$

x	y
−2	
0	
2	

18. $y = 3x - 5$

x	y
0	
2	
3	

19. $y = \frac{2}{3}x + 1$

x	y
−3	
0	
3	

20. $y = \frac{1}{3}x - 4$

x	y
−3	
0	
3	

21. $y = \frac{2}{3}x + 1$

x	y
−4	
−2	
2	

22. $y = \frac{1}{3}x - 4$

x	y
−2	
−1	
4	

23. $-x + 3y = 12$

x	y
−3	
0	
3	

24. $x - 2y = 8$

x	y
−2	
0	
2	

25. $3(x - 1) + 2y = 7 - x$

x	y
0	
2	
4	

26. $4x + 6y = 2(y - x) + 8$

x	y
−2	
0	
2	

27. $y = -2x + 5$

x	y
1	
3	
5	

28. $3x + 2y = 4$

x	y
−2	
0	
4	

29. $y = -3$

x	y
−2	
0	
2	

30. $x = 2$

x	y
	−2
	0
	2

31. $x = 5$

x	y
	−2
	0
	2

32. $y = 0$

x	y
−2	
0	
2	

33. The lines for problems 19 and 21 are identical. Of those two problems, which is easier to work? Why? What can we learn from this?

34. The lines for problems 20 and 22 are identical. Of those two problems, which is easier to work? Why? What can we learn from this?

35. The lines for problems 25 and 27 are identical. Of those two problems, which is easier to work? Why? What can we learn from this?

36. The lines for problems 26 and 28 are identical. Of those two problems, which is easier to work? Why? What can we learn from this?

For the following exercises, graph each linear equation on a Cartesian coordinate system that fits on a 10 x 10 grid. First find three solutions to the equation, then plot the three points, and then finish graphing the line. Don't forget to use arrowheads, label your axes, and show your scale.

37. $y = 2x - 3$

38. $y = x + 2$

39. $y = x$

40. $y = \frac{1}{2}x + 1$

41. $\frac{x + y + 1}{2} = \frac{7}{6}x$

42. $y - 2x = 4x + 3$

43. $4(x + y + 1) = 12$

44. $\frac{y + 3}{4} = \frac{3 - x}{4}$

45. $y = -\frac{2}{3}x$

46. $y = -\frac{3}{2}x + 4$

47. $x = 3$

48. $4x + 3y = -3$

49. $y = -2$

50. $x = 0$

51. $5x + 2y = 10$

52. $y = -3$

53. $x - y = -4$

54. $x + y = 5$

55. $\frac{x}{3} - \frac{y}{4} = \frac{1}{2}$

56. $\frac{1}{4}x + \frac{1}{2}y = \frac{1}{3}$

CHAPTER 6

Proportional Reasoning

What Is a Number?

The world is a complex and dynamic system of interlocking parts. No person or object is completely unaffected by the rest of the world, and the world is not completely unaffected by any individual person or object. Everything is part of a system. When analyzing systems, we use various types of measurement to describe a part of the system with a number value. Because everything is part of a system, we relate things to one another by comparing them.

In this section, we begin to explore how various measurements of different parts of a system are connected to and affected by each other.

6.1 Ratios and Rates

In Chapter 3, we introduced the set of rational numbers. When we use the word "rational" to describe people, ideas, or decisions, we mean that they are reasonable. For example, we might say "in response to the increasing cost of education, the college made a *rational* decision to invest in producing affordable textbooks."

When we use the word "rational" to describe a number, however, it means something quite different. A rational number is a number that is a **ratio** of two integers.

A. Introduction to Ratios

A **ratio** is a specific type of quotient. It compares two numbers and determines their relationship to one another.

> **Ratio**
>
> A quotient that may be used to compare two numbers that represent measurements, amounts, or quantities.

The relationship of the two numbers in a ratio can be expressed as a fraction or in a sentence. For example, since Katie McWilliams is 74 inches tall and Mikayla Pivec is 68 inches tall, we can say that the ratio of Katie's height to Mikayla's height is $\frac{74}{68}$. We can also use the word "to" in place of a fraction bar:

> The ratio of Katie's height to Mikayla's height is 74 to 68.

When we express a ratio as a fraction, that fraction can be simplified. Because $\frac{74}{68}$ and $\frac{37}{34}$ are equivalent fractions, it would also be correct to say that the ratio of Katie's height to Mikayla's height is $\frac{37}{34}$, even though Katie is *not* 37 inches tall, and Mikayla is *not* 34 inches tall.

Example 1

In a recent survey, 750 respondents rated Salem, Oregon's livability as "good" or "excellent," 400 respondents rated Salem's livability as "satisfactory," and 500 respondents rated Salem's livability as "poor." All respondents selected one of these four ratings. Write each ratio below as a simplified fraction:

1. The number of "good" or "excellent" ratings to the number of "satisfactory" ratings.
2. The number of "poor" ratings to the total number of responses.

Solutions

1. In raw numbers, the ratio of "good" or "excellent" ratings to the number of "satisfactory" ratings is $\frac{750}{400}$. This fraction can be simplified to the equivalent fraction $\frac{15}{8}$.

2. The ratio of the number of "poor" ratings to the total number of responses is the quotient of the number of poor ratings over the sum of the total number of ratings: $\frac{500}{750 + 400 + 500} = \frac{500}{1650}$, which reduces to $\frac{10}{33}$.

In situations like this, the ratio is often reported using language such as "10 out of 33 respondents rated Salem's livability as poor." When interpreting such statements, we must keep in mind that the ratio being reported has probably been simplified. There were many more than 33 respondents to the survey, and there were many more than 10 poor ratings. The phrase literally means that 10 out of every 33 people responded with "poor."

Other key phrases that that imply the use of a ratio are:

- x times as many as

- x times more than

- x percent of (the implied ratio is $\frac{x}{100}$)

We can see how these phrases translate mathematically:

- If Jack has 6 times as many bananas as Jill's 3 bananas, then he has $6 \cdot 3 = 18$ bananas.

- If Suresh checks his phone four times a day, and Danielle checks her phone 2.5 times more than Suresh, then Danielle checks her phone $4 \cdot 2.5 = 10$ times a day.

- If 80% of 20 phone calls are robocalls, then $\frac{80}{100} \cdot 20 = \frac{1600}{100} = 16$ calls are robocalls.

Example 2

Rewrite each implied ratio with the phrasing: "the ratio of _____ to _____ is ___ to ___."

1. Florida received two times as much annual precipitation as Oregon.

2. Carlos was 1.2 times faster than Bruce in the men's 5000-meter race.

3. 78% of Rachna's incoming email messages are spam.

Solution

1. The ratio of Florida's annual precipitation to Oregon's annual precipitation is 2 to 1.

2. The ratio of Carlos' speed to Bruce's speed was 12 to 10.

3. The ratio of Rachna's incoming spam email messages to total incoming email messages is 78 to 100.

Note that two of these ratios could be reduced, but it is important to know how to rewrite an implied ratio in its raw numbers, too.

Practice A

When comparing time spent on their MTH 060 homework assignments, Leila reported that she spent 84 minutes, Tom admitted to spending only half as much time as Leila and Bruce had his nose in his book for 56 minutes.

1. What is the ratio of Tom's study time to Leila's study time? Give the reduced ratio.

2. What is the ratio of Bruce's study time to Leila's study time? Give the explicit ratio (without reducing).

3. Again, what is the ratio of Bruce's study time to Leila's study time? This time give the reduced ratio.

4. What is the ratio of Tom's study time to Bruce's study time? Give the explicit ratio (without reducing).

5. What is the ratio of Leila's study time to Bruce's study time? Give the reduced ratio.

6. What is the ratio of Bruce's study time to Tom's study time? Give the reduced ratio.

When you are finished, turn the page to check your work.

B. Rates

When we use a ratio to compare different *types* of measurements, we must include the units with the ratio. This type of ratio is called a **rate**. Here are two examples of rates and the implied ratio:

- June's speed is 65 miles *per* hour → the ratio of distance June traveled to the time June spent traveling is 65 miles to 1 hour.

- Unleaded gas costs 2.79 dollars *per* gallon → the ratio of the amount of money spent to the amount of unleaded gas purchased is $2.79 to 1 gallon.

Notice in both examples the use of the key word, "per." This word almost always signifies a rate. When a rate is reduced such that the second measurement is 1, we call that a **unit rate**. If you drive a car, you may already calculate unit rates: if you drive 90 miles using 3 gallons of gas, the **rate** is 90 miles per 3 gallons of gas, but the **unit rate** is 30 miles per gallon.

The second example is a **unit price**. A **unit price** is the ratio of price to the amount purchased expressed as a rate. Unit prices allow us to compare prices of similar products that are sold in different sized containers. In grocery stores, unit prices are usually posted on the shelf-mounted price tags, as in Figure 1.

Figure 1

Example 3

Figure 1 is the shelf tag for a 7.75 ounce can of chick peas. It shows that the can costs 65 cents. The stated unit price is in dollars per pound. With that in mind, answer these questions.

1. What would the unit price be in cents per ounce?

2. The store also sells 12-ounce cans for 94 cents. What is the unit price in cents per ounce?

3. Which is a better buy (a lower price per ounce), the 7.75 ounce can or the 12-ounce can?

Solution

1. $\frac{65 \text{ cents}}{7.75 \text{ ounces}} \approx 8.39$ cents per ounce

2. $\frac{94 \text{ cents}}{12 \text{ ounces}} \approx 7.83$ cents per ounce

3. The 12-ounce can is a better buy, since each ounce costs slightly less.

Practice B

Roz drives 240 miles in 5 hours while Jace drives 160 miles in 3 hours.

7. What is Roz's speed, in miles per hour?

8. What is Jace's speed, in miles per hour?

9. Who is driving faster, Roz or Jace?

Bank A offers a $10,000 Certificate of Deposit (CD) that returns $10,143.51 in six months, while Bank B offers a $10,000 CD that returns $10,200.73 at the end of nine months.

10. What is the Bank A CD earnings in dollars per month?

11. What is the Bank B CD earnings in dollars per month?

12. If you are simply interested in earnings per month, which is a better deal?

When you're finished, turn the page to check your answers.

C. An Important Ratio in Geometry

A **circle** is defined as the set of points on a plane that are all the same distance from a fixed point. The fixed point is called the **center** of the circle, and the distance from the center to any point on the circle is called the **radius** of the circle. The greatest possible distance between two points on a circle is called the **diameter** of the circle. The distance that one would have to travel along a circle to return to one's starting point without ever reversing direction is called the **circumference** of the circle. There are two ratios involving these measurements that are always true for every circle.

The ratio of diameter to radius is 2 to 1. This fixed ratio gives us the following formula:

> Where r represents the radius of any circle and d represents the diameter of the same circle: $d = 2r$.

The ratio of circumference to diameter is a fixed constant called the **circle ratio** and represented by the Greek letter π, which is spelled "pi" but pronounced like "pie." The exact value of the circle ratio is not a rational number, but its approximate value is $\pi \approx 3.141592654$. This fixed ratio gives us the following formulas:

> Where c represents the circumference, r represents the radius and d represents the diameter of any circle: $c = \pi d$ and $c = 2\pi r$.

The circle ratio also dictates the area of a circle. The ratio of area to the *square* of the radius "r^2" is also π. This fixed ratio gives us the following formula for the area A of a circle with radius r:

> $A = \pi r^2$.

Here is a summary of the formulas that come from these fixed ratios of various measurements of a circle:

> $d = 2r$ (d is diameter, r is radius)
>
> $c = \pi d$ (d is diameter, c is circumference)
>
> $c = 2\pi r$ (c is circumference, r is radius)
>
> $A = \pi r^2$ (A is area, r is radius)

Example 4

Find the following measurements for a pizza that has a diameter of 16 inches:
1. The radius
2. The circumference
3. The area

Solution

1. The radius is 8 inches (solve the equation $16 = 2r$).

2. The circumference is 16π inches, or approximately 50.3 inches.

3. The area is 64π square inches, or approximately 201 square inches.

Practice C

13. Find the radius of a pizza that is 12 inches in diameter.

14. Find the circumference of a pizza that has a 5 inch radius. Give your answer to the nearest tenth of an inch.

15. Find the area of a pizza that has a 6 inch radius. Give your answer to the nearest square inch.

16. Find the diameter of a pizza that is 22 inches in circumference. Give your answer to the nearest inch.

Find the approximate unit price in cents per square inch of area for the following pizzas:

17. 16 inch diameter pizza that costs $14.00

18. 12 inch diameter pizza that costs $9.00.

Which pizza is the better deal? Turn the page to check your answers.

Practice B – Answers

7. Roz's speed is 48 miles per hour.

8. Jace's speed is about 53.3 miles per hour.

9. Jace is driving faster than Roz.

10. The Bank A CD earns approximately $23.92 per month.

11. The Bank B CD earns approximately $22.30 per month.

12. Bank A has the better deal.

Exercises 6.1

For each of the following exercises, write the ratio as a simplified fraction.

1. The approximate number of veterans in Salem, 11,600, to the approximate total population of Salem, 160,000.

2. The approximate number of households in Salem, 57,700, to the approximate total population of Salem, 160,000.

3. The approximate number of persons in poverty in Salem, 30,000, to the approximate number of persons not in poverty in Salem, 130,000.

4. The approximate number of woman owned businesses in Salem, 4400, to the approximate number of men owned businesses in Salem, 6000.

5. In Salem, 11.2% of persons under age 65 years have a disability.

6. In Salem, 11.7% of persons are foreign born.

7. There are approximately 4.5 times as many nonminority owned businesses than minority owned businesses in Salem.

8. There are approximately 3.6 times as many non-Hispanic or Latino residents than Hispanic or Latino residents in Salem.

For the following exercises, rewrite each implied ratio using the syntax "the ratio of _____ to _____ is ___ to ___." For example, given that 40% of animals in a pond are frogs, then we say "The ratio of frogs to total pond animals is 40 to 100."

9. Crater Lake National Park had about 9 times the visitors that Oregon Caves National Monument had in 2016.

10. Lewis and Clark National Historical Park had about 1.3 times the visitors that the John Day Fossil Beds National Monument had in 2016.

11. The John Day Fossil Beds National Monument has about 4.1 times the gross land area of Lewis and Clark National Historical Park.

12. Crater Lake National Park has about 40 times the gross land area of Oregon Caves National Monument.

13. Lewis and Clark National Historical Park had 3% of its annual visitors in December of 2016.

14. Crater Lake National Park had 25% of its annual visitors in July 2016.

15. Oregon Caves National Monument had 18% of its annual traffic count in July of 2016.

16. The Lewis and Clark National Historical Park had 17% of its annual traffic count in July of 2016.

Practice C – Answers

13. The radius is 6 inches.

14. The circumference is 31.4 inches.

15. The area is 113 square inches.

16. The diameter is 7 inches.

17. Unit price for the 16 inch pizza: about 6.96 cents per square inch

18. Unit price for the 12 inch pizza: about 7.96 cents per square inch

The 16 inch pizza is the better deal!

For the following exercises, answer the questions about circles.

17. If the radius of a circular island is about 3 miles, how far would you have to walk in a straight line from a point on the edge through the middle to the opposite side?

18. If the radius of the moon is approximately 1,079 miles, what is the diameter?

19. If the diameter of a CD is 120 millimeters, what is the radius?

20. If a basketball hoop is 18 inches in diameter, what is the radius?

21. How far must a satellite travel to circle around the earth one time on a circular path with a radius of 26,200 mi. (measured from the center of the earth)? Use 3.14 for pi, and give the distance to the nearest thousand miles.

22. A satellite on a circular path with a radius of 26,200 mi. (measured from the center of the earth) will travel around the earth one time in approximately one day. What is the speed of the satellite? Use 3.14 for pi and give the speed to the nearest ten miles per hour.

23. Maxwell would like a round cookie this big. What is the cookie's circumference? Since Maxwell's system of measurement is a bit out of date, assume a diameter of 10 cm and give the circumference to the nearest centimeter.

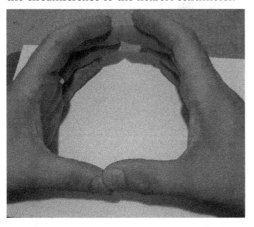

24. Given a circle of circumference 12.1 cm, give the radius to a tenth of a centimeter.

25. Given a circle of circumference 16.3 cm, give the diameter to a tenth of a centimeter.

26. Each of the four clock faces of Big Ben in London is 7.0 meters in diameter. What is the surface area of one of the clock faces? Round your answer to the nearest square meter.

27. Crater lake is roughly a circle with a diameter of 10 km. What is the surface area? Give your answer to the nearest square kilometer.

28. Find the area of a circle with a radius 6.5 ft. Give the area to the nearest square foot.

29. An unruly child ties a chicken leg to a piece of string and jumps onto the dinner table, spinning the chicken leg around his head with a radius of two feet. If the chicken leg circles his head once every half-second, what is the speed of the chicken leg? Give your answer to the nearest foot per second.

30. Find the area of a circle with a diameter 12.1 m. Give the area to the nearest square meter.

31. Find the area of a circle with a circumference of 19.4 mm. Give the area to the nearest square millimeter.

32. Find the area of a circle with a circumference of 14.5 mm. Give the area to the nearest square millimeter.

For the following exercises, calculate the unit rate or unit price.

33. Emma drives 220 miles in four hours. Answer in miles per hour.

34. Liam drives 175 miles in 2.5 hours. Answer in miles per hour.

35. A car with a 17 gallon gas tank can travel 629 miles on one tank. Answer in miles per gallon.

36. A pickup truck with a 21 gallon gas tank can drive 399 miles. Answer in miles per gallon.

37. Olivia can read seven books in four weeks. Answer in books per week.

38. Ethan reads 21 books over the three months of summer. Answer in books per month.

39. A 14 oz can of chili costs $1.39. Answer to the nearest cent per ounce.

40. A 4-pack of cola costs $6. Answer in dollars per can.

41. A rectangular deep dish 9 inch by 14 inch pizza costs $8. Answer to the nearest tenth of a cent per square inch.

42. A rectangular single-serving pizza is 4 inches by 6 inches and costs $2. Answer to the nearest tenth of a cent per square inch.

43. A 14 inch diameter pepperoni pizza cost $5. Using 3.14 for π, answer to the nearest tenth of a cent per square inch.

44. A 14 inch diameter supreme pizza cost $8. Using 3.14 for π, answer to the nearest tenth of a cent per square inch.

For the following exercises, based on unit price, determine which size is a better buy.

45. A 14 ounce can of chili costs $1.39, and a 40 ounce can costs $3.88.

46. A 28 ounce can of baked beans costs $1.58, and a 15 ounce can costs $0.89.

47. A $53.97 bag of lawn fertilizer covers 15,000 square feet, and a $21.47 bag of lawn fertilizer covers 5000 square feet.

48. Garden soil in a 0.75 cubic foot bag costs $4.48, but in a 2 cubic foot bag it costs $7.97.

49. A medium 12 inch diameter pizza sells for $11.99, and a large 14 inch diameter pizza sells for $14.99.

50. A large 14 inch diameter pizza sells for $12, and a family size 16 inch diameter pizza sells for $14.

51. Tubing is stretched across a road and connected to pneumatic car counting devices. Tubing costs $164 for a 200 foot spool, and $246 for a 300 foot spool.

6.2 Solving Proportions

To estimate the trout population in a small remote lake, Travis marked 30 trout by notching their fins and then released the marked fish back into the lake. One week later, he collected a sample of 50 trout from the lake and counted 4 with notched fins. Does this sample provide enough information for the biologist to estimate the trout population in the lake?

This is an example of a scenario that can be modeled with a type of equation called a proportion. A **proportion** is an equation of two ratios. In other words, a proportion is an equation of the form $\frac{a}{b} = \frac{c}{d}$ where a, b, c and d are numbers or variables. Proportions may be used in a number of scenarios involving ratios. In this section, we will learn how to solve proportions and investigate some scenarios that can be modeled with proportions.

A. Solving a Proportion

In Chapter 3, we learned that fractions are equivalent if and only if their cross products are equal. In other words, the proportion $\frac{a}{b} = \frac{c}{d}$ is a true statement if and only if $ad = bc$. This follows from the multiplication principle of equations. If we multiply both sides of $\frac{a}{b} = \frac{c}{d}$ by the non-zero number bd, then the left-hand side of the equation simplifies to ad, and the right-hand side simplifies to bc. We can use this fact to solve a proportion. Here are some examples:

Example 1

Solve each proportion by setting the cross-products equal to each other. Report your answers as decimals.

1. $\dfrac{8}{13} = \dfrac{30}{k}$

2. $\dfrac{t+5}{18} = \dfrac{52}{90}$

3. $\dfrac{24}{2x+2} = \dfrac{44}{4x-3}$

Solutions

1. $\dfrac{8}{13} = \dfrac{30}{k}$

$$\frac{8}{13} = \frac{30}{k}$$

$8k = 13 \cdot 30$ $\qquad\qquad$ Set cross products equal to each other.

$8k = 390$

$\frac{1}{8} \cdot 8k = 390 \cdot \frac{1}{8}$ $\qquad$ Multiply both sides by the reciprocal of 8.

$k = \dfrac{390}{8}$

$k = 48.75$ $\qquad\qquad$ Simplify.

2. $\frac{t+5}{18} = \frac{52}{90}$

$$\frac{t+5}{18} = \frac{52}{90}$$

$90(t+5) = 18 \cdot 52$ Set cross products equal to each other.

$90t + 450 = 936$ Apply Distributive property.

$90t = 486$ Add -450 to both sides.

$t = \frac{486}{90}$ Multiply both sides by the reciprocal of 90.

$t = 5.4$ Simplify.

3. $\frac{24}{2x+2} = \frac{44}{4x-3}$

$$\frac{24}{2x+2} = \frac{44}{4x-3}$$

$24(4x-3) = (2x+2)44$ Set cross products equal to each other.

$96x - 72 = 88x + 88$ Apply Distributive property.

$96x = 88x + 160$ Add 72 to both sides.

$8x = 160$ Add $-88x$ to both sides.

$x = \frac{160}{8}$ Multiply both sides by the reciprocal of 8.

$x = 20$ Simplify.

In each case, the first solution step was setting the cross products equal to each other. This step is often referred to as **cross-multiplication**. After cross-multiplying, we end up with an equation that can be solved using familiar techniques.

Practice A

Solve the following proportions. Then turn the page to check your work. Report your answers as fractions.

1. $\frac{p}{24} = \frac{15}{32}$

2. $\frac{3x}{8} = \frac{7}{4}$

3. $\frac{6}{a} = \frac{14}{15}$

4. $\frac{k+7}{9} = \frac{7}{6}$

5. $\frac{4(r+5)}{5} = \frac{2(r-6)}{8}$

6. $\frac{5}{2n+6} = \frac{3}{n+7}$

B. Inferring Raw Numbers from a Simplified Ratio

Suppose we read in a local high-school newspaper that 6 out of 11 juniors favor their classmate Sherrie to be student body president. We noted in the previous section that it would not be correct to assume that there are only 11 juniors at this high school because the ratio has likely been simplified. In fact, after asking the attendance office, we find that there are 429 juniors. So now the question is how many juniors would vote for Sherrie?

Can we model this scenario with a proportion? Absolutely! We begin by choosing a letter to represent the number of juniors that support Sherrie. Let's choose v for "votes." This variable allows us to write the ratio of Sherrie-supporting juniors to total juniors with raw numbers: $\frac{v}{429}$. We know that this ratio simplifies to $\frac{6}{11}$, so these two different ways to write the same ratio give us a proportion: $\frac{v}{429} = \frac{6}{11}$. When we solve this proportion, we end up with $v = 234$, so there are 234 juniors that will likely vote for Sherrie in the election for student body president. Let's hope she's a good leader.

It's important to remember that ratios reported in the real world are often approximations and not exact values. Approximations are useful because they are easier to understand than exact values. we use the symbol ≈ to indicate that two values are approximately equal. This is very helpful when you need to collect information about something, but you don't have a way to collect every piece of the information, such as a survey of voters or numbers of fish in a lake.

This takes us back to Travis, who is trying to count trout in a remote mountain lake. Travis may use the ratio of marked fish to total fish from his sample (remember, he only caught and notched 30 fish) as an approximation for the ratio of marked fish to total fish in the lake, giving us the approximate proportion $\frac{4}{50} \approx \frac{30}{f}$, where f represents the total number of trout in the lake. Solving this proportion, we find that there are *approximately* 375 trout in the lake.

Approximations may also be used because they communicate information more effectively. For example, suppose that 205,107 of the 256,348 people who are registered to vote in a certain city actually voted in a recent municipal election. The ratio of actual voters to registered voters would probably be reported as *about* 4 out of 5, or 80%, since the exact ratio is *close* to $\frac{4}{5}$. The ratio "4 out of 5" is much easier for most people to digest then the raw number ratio "205,107 to 256,348."

Suppose we only know that about 4 in 5 registered voters cast their ballots and that the total number of people who are registered to vote is 256,348. We could then write the approximate proportion $\frac{n}{256,348} \approx \frac{4}{5}$, where n represents the number of people who voted. Solving this proportion, we find the somewhat mystifying result $n \approx 205,078.4$. We would expect the number of actual voters to be a whole number! However, bear in mind that this is an *approximate* equation. We should probably round our answer to the nearest thousand: about 205,000 people actually voted.

Practice B

An instructor is teaching 3 introductory algebra lecture classes and has a total of 94 students. There are 7 introductory algebra lecture classes at the college.

7. Write an approximate proportion to estimate the number of students in introductory algebra lecture classes at the college.

8. How many students are there in introductory algebra lecture classes at the college? Round your answer to the nearest 10 students.

An instructor lecturing two introductory algebra classes has 19 students getting B's. There are 7 introductory algebra lecture classes at the college.

9. Write an approximate proportion to estimate the number of students in introductory algebra lecture classes that are getting B's at the college.

10. How many introductory algebra students in lecture classes are getting B's at the college?

About 37% of respondents in a recent marketing survey preferred Brand A deodorant to Brand B deodorant. There were a total of 1528 respondents to the survey.

11. Write an approximate proportion to estimate the number of respondents that preferred brand A.

12. How many respondents preferred brand A? Round to the nearest one hundred.

When you are finished, turn the page to check your work.

C. Similarity

When two geometric objects are identical in shape but not identical in size, we say that they are *similar*. Think of a model airplane. A real Cessna 180 (Figure 2) has a wingspan of 10.92 meters and a length of 7.85 meters. A well-made model Cessna 180 will be much smaller, but the ratio of its wingspan to its length should be the same.

Figure 2

Suppose that the model version of this same aircraft has a wingspan of 27 inches. We can use a proportion to determine the appropriate length, l, for the model: $\frac{l}{27} = \frac{7.85}{10.92}$. Approximating our answer to the nearest tenth of an inch, we get $l \approx 19.4$ in. The actual airplane and the model airplane are examples of similar figures. The ratio of any two measurements of the actual airplane is equal to the ratio of the two corresponding measurements of the model airplane.

Because triangles are simpler than other polygons, it is easier to tell whether or not they are similar. Two triangles are similar if and only if their corresponding angles are *congruent* — in other words, identical. Consider the triangles in Figure 3.

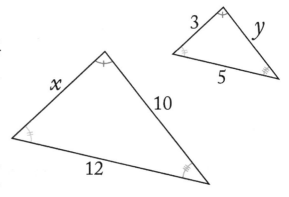

Figure 3

The angles at the left-most vertex of the big triangle and the left-most vertex of the little triangle are congruent. In the same way, the angles at the uppermost vertices of the two triangles are congruent, and the angles at the right-most vertices of the two triangles are also congruent. The ratio of any two side lengths in a triangle will be equal to the ratio of the two corresponding side lengths in a similar triangle, so we can use proportions to calculate side lengths in a similar triangle.

Example 2

1. The triangles below are similar. Find the side lengths labeled x and y.

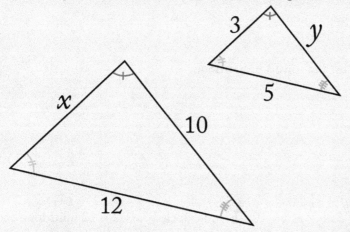

2. The triangles below are similar. Solve for x, and then calculate the side lengths that are labeled in terms of x.

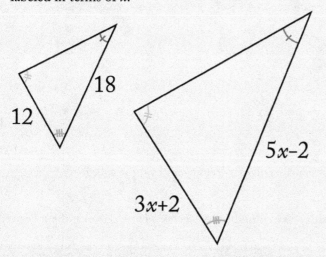

Solutions

1. We use the fact that corresponding sides are opposite corresponding angles to set up and solve the following proportions:

$$\frac{x}{12} = \frac{3}{5}$$
$$x = 7.2$$

And

$$\frac{y}{5} = \frac{10}{12}$$
$$y = \frac{25}{6}$$
$$y \approx 4.17$$

2. Just as we did above, we set up and solve a proportion:

$$\frac{12}{18} = \frac{3x + 2}{5x - 2}$$
$$12(5x - 2) = 18(3x + 2)$$
$$60x - 24 = 54x + 36$$
$$6x = 60$$
$$x = 10$$

Now we evaluate $3x + 2$ and $5x - 2$ at $x = 10$: The side labeled $3x + 2$ is 32 units long and the side labeled $5x - 2$ is 48 units long.

Practice B – Answers

7. $\frac{x}{7} \approx \frac{94}{3}$

8. There are about 220 students in introductory algebra lecture classes at the college.

9. $\frac{x}{7} \approx \frac{19}{2}$

10. Approximately 67 students are getting B's in introductory algebra lecture classes at the college.

11. $\frac{x}{1528} \approx \frac{37}{100}$

12. About 600 respondents preferred Brand A.

Practice C

The triangles below are similar.

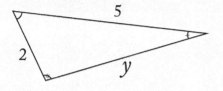

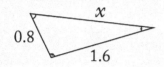

13. Find x

14. Find y

The triangles below are similar.

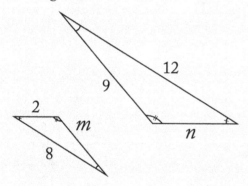

15. Find m

16. Find n

The triangles below are similar.

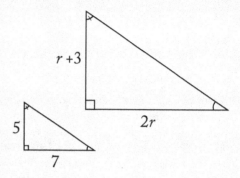

17. Find r

This small triangle is embedded in a larger similar triangle.

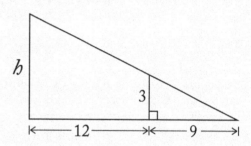

18. Find h

Exercises 6.2

For the following exercises, solve the proportion. Start by setting the cross-products equal to each other.

1. $\frac{2}{3} = \frac{4}{x}$

2. $\frac{3}{2} = \frac{4}{w}$

3. $\frac{3s}{10} = \frac{4}{5}$

4. $\frac{2}{3} = \frac{4v}{9}$

5. $\frac{y}{\frac{3}{4}} = \frac{\frac{2}{3}}{\frac{5}{7}}$

6. $\frac{x}{2.1} = \frac{3.7}{4.8}$

7. $\dfrac{b+3}{16} = \dfrac{9}{4}$

11. $\dfrac{g}{2.1} = \dfrac{5.5}{8.4}$

15. $\dfrac{2}{2n+1} = \dfrac{4}{3n-2}$

8. $\dfrac{t-2}{4} = \dfrac{9}{2}$

12. $\dfrac{2.2}{3.4} = \dfrac{12.1}{d}$

16. $\dfrac{12}{4m-3} = \dfrac{5}{m-3}$

9. $\dfrac{15}{12} = \dfrac{5}{2(x+3)}$

13. $\dfrac{5.1}{m} = \dfrac{3.3}{14.3}$

17. $\dfrac{q+2}{3} = \dfrac{-5q-4}{4}$

10. $\dfrac{3(m-1)}{4} = \dfrac{1}{2}$

14. $\dfrac{6}{5(2x+7)} = \dfrac{3}{2(x+1)}$

18. $\dfrac{3y+2}{4} = \dfrac{5y-2}{5}$

For each of the following exercises, answer the question. Start by setting up a proportion.

19. Yesterday 417 adult Chinook salmon passed the Bonneville Dam, along with 76 Chinook jacks, which are salmon that return to spawn before full grown. If 28,503 jacks have passed the dam this year, estimate the number of adult Chinook salmon that have passed to date this year.

20. Steelhead are a sea-running trout that are similar to salmon. Yesterday 1,365 hatchery-born steelhead passed the Bonneville Dam, along with 699 wild steelhead. If 16,940 hatchery-born steelhead have passed the dam this year, estimate the number of wild steelhead that have passed to date this year.

21. A farmer produced 73,500 pounds of apples last year on his 1,050 tree three-acre farm. If he buys two adjacent acres of land to increase his orchard size, how many pounds of apples per year could he expect once the new trees mature?

22. A 20-acre strawberry farm has about 280,000 plants. Estimate how many plants you could have in a 0.1-acre back yard.

23. If a sample of 40 widgets are tested and 3 are found to be defective, then how many defective widgets would you expect in a production run of 50,000?

24. If six students in a math class of 30 students are getting A's, and if there are about 15,400,000 high school students in math classes in the US, then estimate how many students are getting A's in math in the US.

25. Bobbie has invented a vaccine to prevent people from being attacked by bears, but Dan feels it is better to be attacked by a bear early in life to build up an immunity. In Bobbie's trial, 14 vaccinated participants survived exposure to bears, resulting in a survival rate of 50%. Dan couldn't find anyone willing to test his method. How many test subjects were exposed to bears in Bobbie's trial?

26. Two thirds of the times that Marsha sits down to study her algebra, she accidentally writes a symphony. If Marsha has written 8 symphonies, then how many times has she attempted to study her algebra?

27. Vampires should turn into mosquitoes, not bats. That would make a lot of sense. If two thirds of all mosquitoes are vampires, and if there are 30,000 mosquitoes down by the lake, then how many vampires are there?

28. Herbert sells his inventions to mercenaries. His latest invention is a blindfold. He says "If you can't see the enemy, then the enemy can't see you!" If two out of every three mercenaries visiting his store buy a blindfold, how many mercenaries would he need to visit his store in order to sell 150 blindfolds?

29. Caitlyn asks a vendor if his fish are fresh, and the vendor says "Sure, if you redefine the word "fresh" to mean "ancient" then my fish are very fresh!" Caitlyn then buys 24 salmon and even more trout. If she buys 7 trout for every 6 salmon, then how many trout does she buy?

30. If an introductory algebra class has 8 women for every 7 men, estimate the number of men in a class of 30 students.

31. If a bread recipe calls for 2/3 cup of sugar and 6/5 cups of flour, and you want to increase the recipe to use up your remaining 2 cups of flour, how much sugar should you use?

32. If a recipe calls for a cooking time of 45 minutes for every 2 pounds of beef roast, how many minutes should you cook a roast that weighs 3.5 pounds?

33. A League of Legends support player plays Brand 23 times for every 12 times he plays Zyra. If he has played Zyra 48 times this year, then how many times has he played Brand?

In the following exercises, the paired triangles are similar. Find the missing variable.

34.

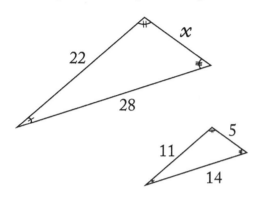

36.

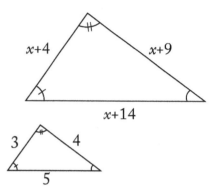

35.

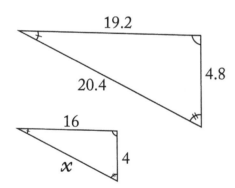

37.

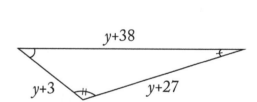

38.

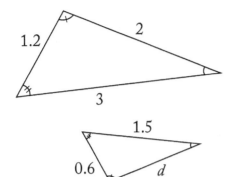

41.

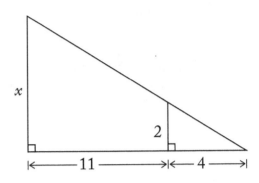

39.

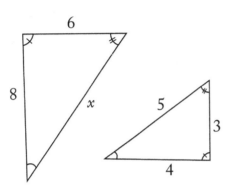

42.

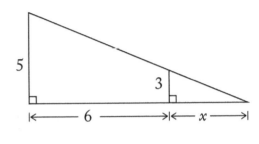

40.

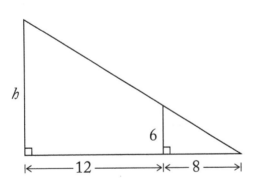

43.

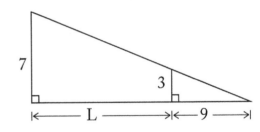

44.

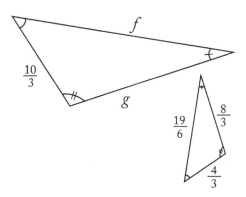

48.

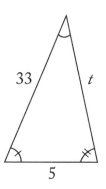
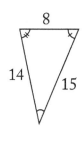

45.

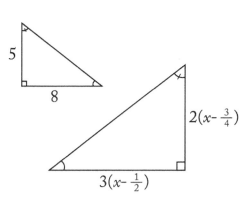

49.

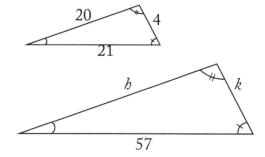

46.

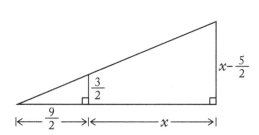

50.

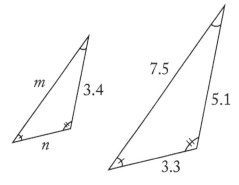

47.

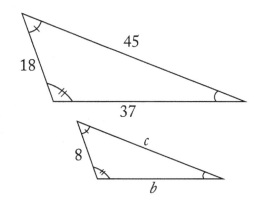

6.3 Dimensional Analysis

In Chapter 4, Simone had just taken up the hobby of rock climbing, and we helped her decide which type of gym membership best suited her needs and budget. Now she is getting ready to venture outdoors, where she plans to climb on actual rocks. However, she must first get the necessary climbing gear. Among her equipment purchases are several bent-gate carabiners, a tool used to link ropes and other equipment together while climbing. While examining the carabiners, she notices some strange symbols stamped onto them (Figure 4).

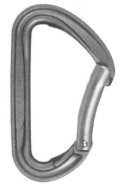

Simone knows enough Physics to know that kN means kilonewtons, a metric measurement of force. She has also learned from reading *Climbing* magazine that the UIAA is the International Climbing and Mountaineering Federation, an organization which

Figure 4

oversees equipment safety standards, so these must be strength ratings. Strength ratings are a measurement of how much weight the carabiners can hold in various configurations.

This information would be more useful to Simone, however, if the weight measurements were in pounds instead of kilonewtons. In this section, we will develop a powerful technique, **dimensional analysis**, for converting measurements in one unit into equivalent measurements in a different unit.

A. Units of Measurement

A **measurement** consists of a number followed by a unit name, such as 300 feet, 101 degrees Fahrenheit, or 23 kilonewtons. The measurement is a numerical description of some physical characteristic of an object. For example, 300 feet is the distance between goal lines on a standard National Football League football field. In this case, the characteristic that is being measured is distance, and the object is a football field.

The unit name represents a standard measurement and is represented by the number 1. For example, when first introduced around the time of the French Revolution, 1 meter was supposed to represent one ten-millionth of the distance from any point on the Earth's equator to the North Pole. According to this standard, a measurement of 400 meters, the distance around the inside lane of a standard running track, would represent 400 ten-millionths of the distance from the equator to the North Pole.

In practice, this was a difficult way to standardize distance measurement. The French Academy of Sciences commissioned a surveying expedition to measure the length of an arc of the prime meridian with unprecedented accuracy so that they could determine the exact length of one meter and then build a platinum rod to serve as a model for the meter. Had this survey been perfect, and if the Earth itself had a uniform shape, then the Earth's circumference, as measured through the North Pole, should be exactly 40,000,000 meters. In fact, the survey was not perfect. It turns out that the actual measurement is closer to 39,931,000 meters — whoops. The point, however, is that there is a physical object, an artifact, which represents the exact length of one meter. Other units of measurement are also tied to a specific artifact.

The United States utilizes two main systems of measurement. The metric system, which has its origins in the original definition of the meter, is the preferred system for applications in the sciences and engineering. It was originally conceived as a universal, comprehensive, and easily used system of measurements that had standards defined in the basic physical properties of the Earth. The US Customary system is also recognized as an official system of measurements in the US, although it is not recognized internationally to the extent of the metric system. Figure 5 summarizes various metric and customary units of measurement, from least to greatest, for different physical characteristics. This table is not comprehensive. Only some of the physical characteristics that can be measured are on this table. For each of those physical characteristics, only some of the units of measurement that can be used are in this table.

Physical characteristic	*Metric units*	*US Customary units*
Distance (or length, width, height, depth, etc.)	Millimeter (mm) Centimeter (cm) Meter (m) Kilometer (km)	Inch (in) Foot (ft) Yard (yd) Fathom Mile (mi)
Area	Square centimeter (cm²) Square meter (m²) Hectare (ha)	Square in (in²) Square foot (ft²) Acre Square mile (mi²)
Volume or Capacity	Milliliter (mL) Liter (L) Cubic meter (m³)	Teaspoon (tsp) Tablespoon (tbsp) Fluid ounce (oz) Cup (c) Pint (pt) Quart (qt) Gallon (gal) Cubic foot (ft³) Cubic yard (yd³)
Force or weight	Newton (N) Kilonewton (kN)	Ounce (oz) Pound (lb) Ton (T)
Mass	Gram (g) Kilogram (kg)	Slug

Figure 5.

Different units of measurement for the same characteristic are called **similar units**. For example, inches and meters both measure length, so they are similar units. In the same way, kilonewtons and pounds both measure force, so they are similar units.

The quotient of two measurements in similar units is always a number; this is an important property of dimensional analysis. Such a quotient always represents the answer to a counting question. For example, an NFL field is 300 feet in length and an iPhone is 14 centimeters in length, so the quotient $\frac{300\,ft}{14\,cm}$ represents the answer to the counting question "how many iPhones can fit end-to-end between the goal lines of a football field?" When we evaluate this quotient, we end up with a simple number: approximately 653.14. There is room for 653 iPhones with a little bit to spare, but not enough room for 654 iPhones. We can represent this with the approximation $\frac{300\,ft}{14\,cm} \approx 653.14$. In the next sections, you'll learn about the importance of converting measurements into similar units.

Example 1

The "kN↔23" stamped onto Simone's carabiners means that each carabiner is rated to hold 23 kN in the direction of its spine (the side of the carabiner opposite the gate) with the gate closed. Simone weighs 105 lbs. Write a counting question that is answered by the quotient $\frac{23\,kN}{105\,lb}$.

Solution:

The quotient $\frac{23\,kN}{105\,lb}$ answers the question "How many Simones can the carabiner safely hold?" Or, since Simone is really just one person, the question should be "how many 105 lb people can the carabiner safely hold?"

Practice A

1. A Fathom is an archaic unit of distance used to measure the depth of a body of water. To make this measurement, a sailor would lower a weight tied to a long rope into the water until it hit bottom, then draw the rope back one arm span at a time, keeping count of arm spans. Each arm span is one fathom.

   Simone uses this same technique to coil her 70 m climbing rope, so that each loop in the coil is the same length as her arm-span, which is 64 in Write a counting question that is answered by the quotient $\frac{70\,m}{64\,in}$.

2. Before emptying a 2-liter bottle of Gatorade into 16.9 oz Water bottles, Rita asks the counting question "How many water bottles will I need?" Write a quotient that will answer this question.

3. Before starting her homework, Sue must read the related section in the book. Otherwise, she probably won't know how to work all of the problems. It takes Sue an average of seven minutes to read each page, and she has allotted two hours for reading time. Write a counting question that is answered by the quotient $\frac{2\,hr}{7\,min}$

4. Caitlyn asks a fish vendor for the average weight of one of his smelt, and the vendor says "The average weight of one of my very fresh smelt is 2.5 ounces." If Caitlyn needs about two pounds of smelt, write a quotient that will answer the counting question "About how many smelt does Caitlyn need?"

5. Rick's property size is 2.4 acres, but only about half of it is lawn (beautiful dandelion-free lawn). If he buys lawn fertilizer bags that each cover 1000 square feet, what counting question is answered by the quotient $\frac{1.2 \text{ acre}}{1000 \text{ ft}^2}$?

6. Jim stores his paperback books on a bookshelf that is 3 feet wide. If the books average 18 mm in thickness, write a quotient that will answer the counting question "About how many books will fit on the shelf?"

When you are finished, turn the page to check your work.

B. Equal Measurements and Unit Fractions

Two measurements are equal if both accurately measure the same physical characteristic of the same object. If either measurement is approximately correct, then the measurements are approximately equal. For example, the distance between goal lines in a standard NFL field can be measured as 300 ft or as 100 yd. This means that 300 ft and 100 yd are equal measurements. We can represent this mathematically with the equation 300 ft = 100 yd. In a similar way, a standard soda can measures 12.5 oz and about 355 mL in capacity. This means that the measurements 12.5 oz and 355 are approximately equal. We can represent this mathematically with the approximation $12.5 \text{ oz} \approx 355 \text{ mL}$.

Let's use these examples to think about what happens when we evaluate the quotient of two equal measurements. The quotient $\frac{300 \text{ ft}}{100 \text{ yd}}$ answers the question "how many football fields will fit on one football field?" Similarly, the quotient $\frac{12.5 \text{ oz}}{355 \text{ mL}}$ answers the question "how many approximate soda cans will fit in one soda can?" You should see that the answer to the first question is exactly one and the answer to the second question is approximately one.

The first quotient is an example of a **unit fraction**, and the second is an **approximate unit fraction**. A unit fraction can also be called a *conversion* fraction.

Unit Fractions and Approximate Unit Fractions

A unit fraction is a quotient of equal measurements. The value of a unit fraction is exactly equal to one.

An approximate unit fraction is a quotient of approximately equal measurements. The value of an approximate unit fraction is approximately equal to one.

Every pair of equal measurements yields two unit fractions, which are reciprocals of each other. For example, both unit fractions $\frac{300\,ft}{100\,yd}$ and $\frac{100\,yd}{300\,ft}$ come from the equation 300 ft = 100 yd. The following table (Figure 6) summarizes many useful equal or approximately equal measurements, each of which yields two unit fractions:

Physical Characteristic	Metric to Metric	Customary to Customary	Metric to Customary
Distance	1 m = 1000 mm 1 m = 100 cm 1 km = 1000 m 1 cm = 10 mm	1 ft = 12 in 1 yd = 3 ft 1 mi = 5280 ft	1 in ≈ 2.54 cm 1 m ≈ 39.37 in 1 mi ≈ 1.609 km
Area	1 ha = 10,000 m²	1 acre = 43,560 ft²	1 ha ≈ 2.471 acres
Volume	1 mL = 1 cm³ 1 L = 1000 mL 1 m³ = 1000 L	1 tbsp = 3 tsp 1 oz ≈ 2 tbsp 1 c = 8 oz 1 pt = 2 c 1 qt = 2 pt 1 gal = 4 qt 1 ft³ ≈ 7.48 gal 1 yd³ = 27 ft³	1 pt ≈ 473.2 cm³ 1 L ≈ 1.057 qt 1 tsp ≈ 5 mL 1 ft³ ≈ 28.32 L
Force	1 kN = 1000 N	1 lb = 16 oz 1 T = 2000 lb	1 lb ≈ 4.448 N
Mass	1 kg = 1000 g		1 slug ≈ 14.59 kg

Figure 6

Example 2

1. Write two unit fractions involving quarts and gallons.

2. Write two approximate unit fractions involving miles and kilometers.

Solution

1. $\frac{4\,qt}{1\,gal}$ and $\frac{1\,gal}{4\,qt}$

2. $\frac{1.609\,km}{1\,mi}$ and $\frac{1\,mi}{1.609\,km}$

Practice B

7. Write two unit fractions involving square feet and acres.

8. Write two approximate unit fractions involving pounds and Newtons.

9. Write two unit fractions involving inches and centimeters.

10. Write two approximate unit fractions involving cubic feet and liters.

11. Write two unit fractions involving grams and kilograms.

12. Write two approximate unit fractions involving meters and inches.

When you are finished, turn the page to check your work.

C. Converting Measurements into Similar Units

Occasionally, we may wish to express a measurement in units that are different from those originally used. In the introduction to this section, Simone found herself in such a position, wanting to know the strength of her carabiners. We will return to Simone shortly, but let us begin with a different example.

Suppose that Ben is training to run a marathon. His training regimen gives distances for his training runs in kilometers. The standard length of a marathon is 26.2 miles. Ben would like to convert this distance to kilometers so that he knows how it compares to his training runs. He knows that the unit fractions $\frac{1.609\,\text{km}}{1\,\text{mi}}$ and $\frac{1\,\text{mi}}{1.609\,\text{km}}$ both involve miles and kilometers, and that unit fractions are equal in value to the number 1, so he correctly figures that multiplying a measurement by a unit fraction does not change the measurement:

$$26.2 \text{ mi} \approx 26.2 \text{ mi} \cdot \frac{1.609\,\text{km}}{1\,\text{mi}}$$
$$\approx \frac{26.2\,\cancel{\text{mi}} \cdot 1.609\,\text{km}}{1\,\cancel{\text{mi}}}$$
$$\approx \frac{26.2 \cdot 1.609\,\text{km}}{1}$$
$$\approx 42.2\,\text{km}$$

In the second step we canceled the unit name "mi" in the numerator and the denominator as if we were reducing a fraction by canceling common factors. In fact, that is exactly what we were doing!

Notice that this conversion is an approximation since we are multiplying by an approximate unit fraction. Notice also that there are two unit fractions involving miles and kilometers. It's very important for Ben to select the correct one! Suppose Ben tried multiplying by the other unit fraction: $26.2 \text{ mi} \approx 26.2 \text{ mi} \cdot \frac{1\,\text{mi}}{1.609\,\text{km}}$. This approximation is mathematically correct but it is not helpful since the miles do not cancel!

Sometime you may need to multiply by more than one unit fraction. Think back to Simone's 70 m climbing rope. Since she needs to be able to return to the ground after climbing a route, the rope must be at least twice as long as the route that she is climbing. That means that she can only climb routes that are not longer than 35 m. If she wants to climb a route that is 130 feet, is her rope long enough?

The table of equal and approximately equal measurements in Figure 6 doesn't have an equation involving feet and meters, but it does have the following two equations: $1\,\text{ft} = 12\,\text{in}$ and $1\,\text{m} \approx 39.37\,\text{in}$ Simone should start by converting 130 feet into inches, and and then convert the measurement from inches into meters. She can do so in a single step:

$$130\,\text{ft} \approx 130\,\text{ft} \cdot \frac{12\,\text{in}}{1\,\text{ft}} \cdot \frac{1\,\text{m}}{39.37\,\text{in}}$$

$$\approx \frac{130\,\text{ft} \cdot 12\,\text{in} \cdot 1\,\text{m}}{1\,\text{ft} \cdot 39.37\,\text{in}}$$

$$\approx \frac{130 \cdot 12 \cdot 1}{1 \cdot 39.37}\,\text{m}$$

$$\approx 39.6\,\text{m}$$

It looks like this route is too long for Simone to climb with her 70 m rope. She should befriend a climber with an 80 m rope that she can borrow!

We can also use the technique of multiplying by unit fractions to simplify quotients of similar measurements. For example, the question "how many 2 L bottles of soda can be poured into a 5 gal cooler?" is answered by the quotient $\frac{5\,\text{gal}}{2\,\text{L}}$. However, this answer is not very useful when we're standing in the soda aisle at the grocery store, trying to decide how many bottles to buy! Using the equations $1\,\text{L} \approx 1.057\,\text{qt}$ and $1\,\text{gal} = 4\,\text{qt}$ we simplify as follows:

$$\frac{5\,\text{gal}}{2\,\text{L}} \approx \frac{5\,\text{gal}}{2\,\text{L}} \cdot \frac{1\,\text{L}}{1.057\,\text{qt}} \cdot \frac{4\,\text{qt}}{1\,\text{gal}}$$

$$\approx \frac{5\,\text{gal} \cdot 1\,\text{L} \cdot 4\,\text{qt}}{2\,\text{L} \cdot 1.057\,\text{qt} \cdot 1\,\text{gal}}$$

$$\approx \frac{5 \cdot 1 \cdot 4}{2 \cdot 1.057 \cdot 1}$$

$$\approx 9.46$$

It looks like 9 bottles will almost fill the cooler. However, if we wish to fill the cooler right to the top we should buy 10 bottles.

Practice A – Answers

1. The quotient $\frac{70\,\text{m}}{64\,\text{in}}$ answers the question "How many loops are in Simone's coiled rope?"

2. The quotient $\frac{2\,\text{L}}{16.9\,\text{oz}}$ answers the question, "How many water bottles will I need?"

3. The quotient $\frac{2\,\text{hr}}{7\,\text{min}}$ answers the question, "About how many pages does Sue have time to read?"

4. The quotient $\frac{2\,\text{lb}}{2.5\,\text{oz}}$ answers the question, "About how many smelt does Caitlyn need?"

5. The quotient $\frac{1.2\,\text{acre}}{1000\,\text{ft}^2}$ answers the question, "How many bags of lawn fertilizer should Rick buy?"

6. The quotient $\frac{3\,\text{ft}}{18\,\text{mm}}$ answers the question, "About how many books will fit on the shelf?"

Looking back, when we found that 653 iPhones fit end-to-end in a NFL football field, we only could do that after we converted the 300 ft of the football field into 9144 centimeters to determine how many 14 cm fit into its length. Many questions like these can be answered if you know how to convert different types of units!

Practice C

Use the technique of multiplying by unit fractions to do the following:

13. Convert 15 inches to centimeters.

14. Simplify the quotient $\frac{2 \text{ lb}}{2.5 \text{ oz}}$

15. Convert 13 feet to centimeters.

16. Simplify the quotient $\frac{300 \text{ ft}}{14 \text{ cm}}$. Give your answer to the nearest whole number.

17. Convert 23 kilonewtons into pounds. Give your answer to the nearest pound.

18. Use the technique of multiplying by unit fractions to simplify the quotient $\frac{23 \text{ kN}}{105 \text{ lb}}$. Give your answer to the hundredths.

When you are finished, turn the page to check your work.

Exercises 6.3

Write a quotient that will answer the counting question.

1. How many 40 foot maple trees would need to be stacked vertically to reach the height of the top of Mount Kilimanjaro, 5890 meters high?

2. How many 40 foot maple trees would need to be stacked to reach the moon (approximate distance to the moon 2.389×10^5 miles)?

3. How many 40 foot maple trees would need to be stacked to reach the sun (approximate distance to the sun 9.296×10^7 miles)?

4. How many 6 ft people laying end-on-end would it take to stretch around the equator (approximate distance around the equator 40,000 kilometers)?

5. How many 0.4 oz shrimp are needed for a recipe that calls for one pound?

6. How many 60 lb kegs of home brew can you carry in a half-ton pickup?

7. Stan is making his own home-made weightlifting equipment. If he fills a bucket with 3 ounce fishing weights, how many fishing weights does he need to end up with 80 pounds or more (neglecting the weight of the bucket)?

8. Maxwell buys 3 pounds of cookies to eat each month. If each cookie weighs 2 ounces, how many cookies does he eat each month?

Write two unit fractions involving each set of two units.

9.   Pounds and ounces

10. Tons and pounds

11. Cubic centimeters and milliliters

12. Liters and milliliters

13. Miles and feet

14. Slugs and kilograms

15. Meters and centimeters

16. Centimeters and millimeters

Use the technique of multiplying by unit fractions to simplify the following quotients. Give your answer to the nearest whole number.

17. $\dfrac{0.12 \text{ acre}}{1000 \text{ ft}^2}$

18. $\dfrac{2.72 \text{ m}}{21.51 \text{ in}}$

19. $\dfrac{1 \text{ lb}}{0.4 \text{ oz}}$

20. $\dfrac{47 \text{ acres}}{59,560 \text{ m}^2}$

21. $\dfrac{3 \text{ ft}}{18 \text{ mm}}$

22. $\dfrac{324 \text{ m}}{40 \text{ ft}}$

23. $\dfrac{2 \text{ L}}{16.9 \text{ oz}}$

24. $\dfrac{80 \text{ lb}}{3 \text{ oz}}$

Use the technique of multiplying by unit fractions to perform the following conversions. Give your answers to the tenth of a unit.

25. 9.30 in to cm

26. 17.3 in to cm

27. 49.7 cm to in

28. 2.52 lb to oz

29. 3.20 lb to oz

30. 14 cm to in

31. 4.35 cups to oz

32. 3.0 mi to km

33. 75 mm to cm

34. 10.2 lb to N

35. 15 gal to ft³

36. 38.6 L to gal

37. 1.8 m to ft

38. 9300 m to mi

39. 50 c to L

40. 2500 mL to qt

41. 2.3 gal to L

42. 935.0 oz to N

43. 1000.5 N to oz

44. 0.02513 slugs to g

45. 90,000 g to slugs

Simplify or convert the following, presenting your answers in scientific notation with the indicated number of significant digits.

46. Simplify $\dfrac{5980 \text{ m}}{40 \text{ ft}}$

    (approximate number of maple trees that stacked vertically would reach the top of Mount Kilimanjaro: use 3 significant figures).

47. Simplify $\dfrac{9.296 \times 10^7 \text{ mi}}{40 \text{ ft}}$

    (approximate number of maple trees stacked to reach the sun: use 4 significant figures).

48. Convert 47 acres to square feet (Alcatraz Island area: use 2 significant figures).

49. Convert 2.389×10^5 miles to meters (approximate distance to the moon: use 4 significant figures).

50. Convert 9.296×10^7 miles to kilometers (approximate distance to the sun: use 4 significant figures).

Practice C – Answers

13. $15 \text{ in} = 15 \text{ in} \left(\dfrac{2.54 \text{ cm}}{1 \text{ in}} \right) = 38.1 \text{ cm}$

14. $\dfrac{2 \text{ lb}}{2.5 \text{ oz}} = \dfrac{2 \text{ lb}}{2.5 \text{ oz}} \left(\dfrac{16 \text{ oz}}{1 \text{ lb}} \right) = 12.8$

15. $13 \text{ ft} = 13 \text{ ft} \left(\dfrac{12 \text{ in}}{1 \text{ ft}} \right) \left(\dfrac{2.54 \text{ cm}}{1 \text{ in}} \right) = 396.24 \text{ cm}$

16. $\dfrac{300 \text{ ft}}{14 \text{ cm}} = \dfrac{300 \text{ ft}}{14 \text{ cm}} \left(\dfrac{2.54 \text{ cm}}{1 \text{ in}} \right) \left(\dfrac{12 \text{ in}}{1 \text{ ft}} \right) \approx 653$

17. $23 \text{ kN} \approx 23 \text{ kN} \cdot \dfrac{1000 \text{ N}}{1 \text{ kN}} \cdot \dfrac{1 \text{ lb}}{4.448 \text{ N}} \approx 5171 \text{ lb}$

18. $\dfrac{23 \text{ kN}}{105 \text{ lb}} \approx \dfrac{23 \text{ kN}}{105 \text{ lb}} \cdot \dfrac{1000 \text{ N}}{1 \text{ kN}} \cdot \dfrac{1 \text{ lb}}{4.448 \text{ N}} \approx 49.25$

Solutions to Odd-Numbered Exercises

Chapter 1: Whole Numbers

1.1 Addition and Subtraction

1. 11
3. 11
5. 59
7. 132
9. 1,068
11. The commutative property of addition guarantees that the answers to questions 1 and 3 are the same.
13. 25
15. 25
17. 54

19. 256
21. The associative property of addition guarantees that the answers to questions 13 and 15 are the same.
23. 30
25. 95
27. 175
29. 207
31. $36 = 25 + 11$
33. $478 = 261 + 217$
35. $887 = 38 + 849$

37. $7,611 = 4,653 + 2958$
39. 5
41. 0
43. Not a whole number
45. 16
47. 344
49. 5
51. 5
53. 4
55. 28

1.2 Solving Equations

1. is a solution
3. is not a solution
5. is a solution
7. is not a solution
9. is a solution
11. 19
13. 5
15. 2
17. 758
19. 267
21. $v - 8 = 13$
    $v = 8 + 13$
    $v = 13 + 8$
    $v - 13 = 8$

23. $16 - r = 7$
    $16 = r + 7$
    $16 = 7 + r$
    $16 - 7 = r$
25. $25 - x = 9$
    $25 = x + 9$
    $25 = 9 + x$
    $25 - 9 = x$
27. $113 - s = 85$
    $113 = s + 85$
    $113 = 85 + s$
    $113 - 85 = s$
29. 21
31. 9
33. 16

35. 28
37. 237
39. 37
41. 35
43. 35
45. 255
47. 503
49. 36

1.3 Multiplication and Division

1. 28

3. 56

5. 8

7. 56

9. 0

11. 11 + 11 + 11 + 11 + 11 + 11

13. $x + x + x + x$

15. 6 + 6

17. 9

19. 1 + 1 + 1 + 1 + 1 + 1 + 1 + 1 + 1

21. $2 \cdot 2 \cdot 2 \cdot 2 \cdot 2$

23. $11 \cdot 11 \cdot 11 \cdot 11 \cdot 11 \cdot 11$

25. $c \cdot c \cdot c$

27. $1 \cdot 1 \cdot 1 \cdot 1 \cdot 1 \cdot 1 \cdot 1 \cdot 1 \cdot 1$

29. 9

31. 3

33. not a whole number

35. 0

37. undefined

39. 7

41. There is a case where a^b and b^a have the same value when a and b are different whole numbers, but there are many cases where they do not. If a and b are 2 and 4, then a^b and b^a have the same value. Answers vary for cases where a^b and b^a do not have the same value.

43. In expanded form 0^3 is $0 \cdot 0 \cdot 0$. The value of 0^3 is 0 when a is greater than zero. It does not matter what number a represents, as long as a is greater than zero.

45. 1

47. 12

49. 48

51. 8

53. 16

55. William will need nine trays. There will be 16 tumblers on the last tray.

57. a) 100
b) 1,000
c) 10,000

1.4 Order of Operations

1. 3(4)

3. $12 \cdot 6$

5. $4 \cdot 3$

7. 18 − 12

9. 2^3

11. 5^4

13. 5 − 5

15. 24 ÷ 6

17. 3(4)

19. $5 \cdot 2$

21. $2 \cdot 3$

23. 4 − 3

25. 14

27. 13

29. 144

31. 11

33. 4,096

35. 621

37. undefined

39. 8

41. 4,096

43. 2

45. 50

47. 3

49. undefined

51. 4

53. 18

55. 3

57. 36

59. No. Enter $(7 + 8) \div (4 - 1)$

1.5 Solving Equations

1. $\frac{40}{p} = 5$

   $5p = 40$

   $p \cdot 5 = 40$

   $\frac{40}{5} = p$

3. $\frac{65}{z} = 5$

   $5z = 65$

   $z \cdot 5 = 65$

   $\frac{65}{5} = z$

5. $m - 28 = 53$

   $m = 28 + 53$

   $m = 53 + 28$

   $m - 53 = 28$

7. $23 - 12 = b$

   $23 - 12 = b$

   $23 = b + 12$

   $23 - b = 12$

9. $\frac{u}{3} = 5$

   $24 \cdot 3 = u$

   $3 \cdot 24 = u$

   $\frac{u}{24} = 3$

11. $\frac{36}{u} = 5$

    $6u = 36$

    $u \cdot 6 = 36$

    $\frac{36}{6} = u$

13. $62 - 21 = b$

    $62 = 21 + b$

    $62 = b + 21$

    $62 - b = 21$

15. $\frac{f}{7} = 20$

    $20 \cdot 7 = f$

    $7 \cdot 20 = f$

    $\frac{f}{20} = 7$

17. $\frac{93}{k} = 3$

    $3k = 93$

    $k \cdot 3 = 93$

    $\frac{93}{3} = k$

19. $512 - z = 80$

    $512 = z + 80$

    $512 = 80 + z$

    $512 - 80 = z$

21. $\frac{c}{110} = 6$

    $6 \cdot 110 = c$

    $110 \cdot 6 = c$

    $\frac{c}{6} = 110$

23. $621 - 396 = w$

    $621 = 396 + w$

    $621 = w + 396$

    $621 - w = 396$

25. 8

27. 13

29. 81

31. 11

33. 72

35. 6

37. 41

39. 140

41. 31

43. 432

45. 660

47. 225

49. 1150

Chapter 2: Integers

2.1 Introduction to Integers

1. $3 > -8$

3. $3 < 8$

5. $-9 < -5$

7. $0 > -2$

9. $1 > -1$

11. $-5 < 3$

13. $-5 < -(-8)$

15. $1 > -1$

17. $0 > -3$

19. -3 is read as "negative three."

21. $-(-8)$ is read as "opposite of negative eight."

23. $|-7|$ is read as "absolute value of negative seven."

25. $-|x|$ is read as "opposite of absolute value of x."

27. $-w$ is read as "opposite of w."

29. 5

31. 4

33. -12

35. -3

37. -8

39. 0

41. 0

43. 5

45. 9

47. 13

49. 6

2.2 Addition and Subtraction

1. $-5 + (-9)$
3. $-23 + (-22)$
5. $4 + 7$
7. $4 + 35$
9. $-11 + 6$
11. $-6 + 17$
13. $5 + (-13) + 3$
15. -4
17. -18
19. 6

21. 14
23. -40
25. 12
27. -372
29. -29
31. -12
33. -4
35. -44
37. 120
39. -12

41. -29
43. 30
45. Bob's net worth is $-\$40,500$
47. Yau's net worth is $-\$31$
49. 7
51. 7
53. The value of $|a - b|$ represents the distance between a and b on a number line.
55. 1010

2.3 Multiplication and Division

1. -20
3. 6
5. -6
7. -3
9. undefined
11. -49
13. 24
15. 24
17. 0

19. 0
21. 42
23. -24
25. -240
27. 1
29. 9
31. -32
33. 81
35. -16

37. -81
39. 20
41. -4
43. 5
45. -24
47. 2
49. 0
51. 0
53. undefined

2.4 Solving Equations

1. 17
3. -21
5. 14
7. 35
9. -15
11. -25
13. 7
15. The addition must be undone first. The equivalent equation that results from adding negative eight to each side of the original equation is $4c = 24$

17. The addition must be undone first. The equivalent equation that results from adding 17 to each side of the original equation is $21b = 0$
19. The addition must be undone first. The equivalent equation that results from adding negative eight to each side of the original equation is $9d = 27$
21. The addition must be undone first. The equivalent equation that results from adding negative 93 to each side of the original equation is $-54 = 54h$

23. The addition must be undone first. The equivalent equation that results from adding 15 to each side of the original equation is $-x = -9$
25. The multlipication must be undone first. The equation from the fact family that undoes this operation is $\frac{161}{7} = r + 14$
27. The multlipication must be undone first. The equation from the fact family that undoes this operation is $\frac{-99}{3} = 23 + v$

29. The division must be undone first. The equations from the fact family that undo this operation are
$q - 23 = 14 \cdot 6$ and $q - 23 = 6 \cdot 14$

31. The division must be undone first. The equation from the fact family that undoes this operation is $\frac{22}{2} = 4 + a$

33. 6

35. 0

37. 3

39. -1

41. 9

43. 9

45. -56

47. 107

49. 7

51. 11

53. -87

55. 8

57. 13

59. 27

Chapter 3: Rational Numbers

3.1 Introduction to Rational Numbers

1. 9

3. 6

5. The rational number $\frac{37}{5}$ is between 7 and 8.

7. The rational number $\frac{58}{7}$ is between 8 and 9.

9. −6

11. −6

13. 0

15. The rational number $\frac{42}{-32}$ is between −1 and −2.

17. The rational number $-\frac{32}{50}$ is between 0 and −1.

19. $\frac{34}{17}$

21. $-\frac{2}{3}$

23. $-\frac{23}{26}$

25. $-\frac{2}{3}$

27. $\frac{4}{10}$

29. $-\frac{24}{100}$

31. $\frac{237}{100}$

33. $-\frac{52}{10}$

35. $\frac{3}{1000}$

37. $\frac{3}{8}$

39. $\frac{13}{16}$

41. $\frac{4}{3}$

43. 0.9

45. 0.1

47. −0.019

49. −2.3

51. −23.1

53. 0.1233

55. 0.0001

57. −83.5

3.2 Divisibility

1. 1, 2, 3, 6, and 9

3. 1

5. 1, 3, and 5

7. 1, 2, 3, 4, 6, and 8

9. 1, 2, 4, 5, and 10

11. 1

13. 1, 2, 3, 4, 5, 6, 8, 9, and 10

15. 1, 2, 4, 5, 8, and 10

17. 1, 2, 3, 4, 6, and 9

19. 1, 5, and 7

21. 1, 2, 3, and 6

23. 1, 2, 7, and 14

25. 1, 2, 3, 4, 6, 8, 12, 16, 24, and 48

27. 1 and 67

29. 1, 2, 41, and 82

31. 1, 2, 3, 4, 6, 7, 12, 14, 21, 28, 42, and 84

33. 1, 3, 9, 11, 33, and 99

35. 2

37. 6

39. 1

41. 2

43. 2

45. 14

47. 6

49. 6

51. 2

53. 24

55. 14

57. 36

59. 36

61. 240

63. 120

65. Answers may vary. Here is one possible answer. Let n be an integer. Subtract the ones digit from the remaining digits. If the answer is divisible by 11 then so is n itself. You can use this test on the answer to find out if it is divisible by 11. For example, is 1507 divisible by 11? Calculate $150 - 7 = 143$. Is 143 divisible by 11? Calculate $14 - 3$. We know that 11 is divisible by 11, therefore 143 is divisible by 11, and therefore 1507 is divisible by 11. Also for example, is 3679 divisible by 11? Calculate $367 - 9 = 358$. Is 358 divisible by 11? Calculate $35 - 8$. We know that 27 is not divisible by 11, therefore 358 is not divisible by 11, and therefore 3679 is not divisible by 11.

67. Answers may vary. Here is one possible answer. Let n be an integer. Subtract 5 times the ones digit from the remaining digits. If the answer is divisible by 17 then so is n itself. You can use this test on the answer to find out if it is divisible by 17. For example, is 1507 divisible by 17? Calculate $150 - 5 \cdot 7 = 115$. Is 115 divisible by 17? Calculate $11 - 5 \cdot 5 = -14$. We know that -14 is not divisible by 17, therefore 115 is not divisible by 17, and therefore 1507 is not divisible by 17. Also for example, is 1751 divisible by 17? Calculate $175 - 5 \cdot 1 = 170$. Is 170 divisible by 17? Calculate $17 - 5 \cdot 0$. We know that 17 is divisible by 17, therefore 170 is divisible by 17, and therefore 1751 is divisible by 17.

3.3 Prime Factorization

1. Prime
3. Prime
5. Composite, 2
7. Composite, 3 or 5
9. Prime
11. Composite, 7 or 43
13. Composite, 11 or 23
15. Prime
17. Prime
19. Composite, 19
21. $2 \cdot 11$

23. $3^2 \cdot 7$
25. $2^2 \cdot 11$
27. $5 \cdot 7 \cdot 11$
29. $3^3 \cdot 5$
31. $2^3 \cdot 5 \cdot 13^2$
33. $2^3 \cdot 7^3$
35. $3 \cdot 5^3 \cdot 11$
37. GCF 9; LCM 945
39. GCF 22; LCM 44
41. GCF 1; LCM 2772
43. GCF 11; LCM 1540

45. GCF 7; LCM 3465
47. GCF 55; LCM 28,875
49. GCF 4; LCM 30,184
51. GCF 5; LCM 5,577,000
53. GCF 1; LCM 221
55. GCF 1; LCM 54,991
57. The LCM is their product.
59. Nothing, because $\text{GCF}(a, b) \cdot \text{lcm}(a, b) = a \cdot b$ is true for all a and b.
61. Yes, the technique works for three or more numbers.
63. a

3.4 Equivalent Fractions

1. $\frac{2}{3}$
3. $\frac{2}{3}$
5. $\frac{3}{2}$
7. $\frac{4}{3}$
9. $-\frac{3}{4}$
11. Already reduced
13. $-\frac{25}{16}$
15. $-\frac{1}{4}$
17. $\frac{9}{2}$

19. $\frac{17}{4}$
21. Already reduced
23. Equivalent
25. Not Equivalent
27. Equivalent
29. Equivalent
31. Not Equivalent
33. Equivalent
35. Equivalent

37. $\frac{12}{27}$
39. $\frac{-4}{10}$
41. No such fraction exsits
43. $\frac{12}{39}$
45. $\frac{16}{-12}$
47. $\frac{36}{16}$
49. No such fraction exsits
51. $\frac{-27}{45}$

3.5 Multiplication and Division

1. $\frac{5}{12}$

3. $\frac{3}{4}$

5. $-\frac{10}{21}$

7. $\frac{27}{16}$

9. $\frac{6}{5}$

11. $-\frac{12}{5}$

13. $-\frac{8}{5}$

15. $\frac{18}{5}$

17. $\frac{20}{77}$

19. $\frac{1}{55}$

21. $\frac{3}{2}$

23. $-\frac{1}{80}$

25. 0

27. Reciprocals

29. Not reciprocals

31. Not reciprocals

33. Reciprocals

35. Reciprocals

37. Not reciprocals

39. $\frac{1}{6}$

41. $\frac{9}{2}$

43. $\frac{21}{20}$

45. $\frac{10}{9}$

47. 0

49. $\frac{4}{21}$

51. $-\frac{4}{9}$

53. It will take Monica $\frac{125}{12}$ hours to finish typing her paper.

55. Bryce will need $\frac{3}{2}$ packages of seed.

57. James supplied gravel for 16 fish tanks.

59. Six students ate pizza.

61. The deer is $\frac{8}{3}$ old.

3.6 Comparing Rational Numbers

1. d or e, $\frac{2}{9} < \frac{4}{7}$

3. b, $\frac{-2}{9} < \frac{1}{7}$

5. a, $\frac{7}{9} > \frac{4}{9}$

7. c or e, $\frac{-9}{8} < \frac{-4}{7}$

9. e, $-\frac{12}{5} > -\frac{17}{7}$

11. c or e, $\frac{8}{9} < \frac{14}{11}$

13. e, $\frac{5}{8} > \frac{4}{7}$

15. a, $-\frac{3}{2} > -\frac{5}{2}$

17. d or e, $-\frac{4}{7} > -\frac{5}{6}$

19. d or e, $-\frac{8}{7} > -\frac{9}{5}$

21. e, $\frac{7}{-9} < \frac{5}{-9}$

23. $\frac{3}{4} > \frac{5}{8}$

25. $\frac{3}{4} < \frac{9}{11}$

27. $\frac{4}{3} < \frac{7}{5}$

29. $\frac{7}{3} > \frac{5}{4}$

31. $\frac{8}{5} < \frac{11}{61}$

33. $\frac{11}{15} < \frac{7}{9}$

35. $\frac{11}{15} < \frac{7}{9}$

37. $\frac{11}{6} > \frac{16}{9}$

39. $\frac{-3}{4} < \frac{5}{-7}$

41. $-\frac{4}{3} > -\frac{8}{5}$

43. $\frac{-11}{-5} < \frac{-19}{-8}$

45. $\frac{9}{16} > \frac{-7}{12}$

47. $-\frac{2}{3}, \frac{13}{17}, \frac{4}{3}$

49. $\frac{4}{9}, \frac{5}{8}, \frac{8}{5}$

51. $\frac{13}{17}, \frac{4}{3}, \frac{5}{2}$

53. $-\frac{11}{6}, -\frac{3}{4}, \frac{2}{3}$

55. $\frac{7}{11}, \frac{2}{3}, \frac{4}{3}$

57. $-\frac{6}{13}, -\frac{3}{7}, -\frac{5}{13}$

59. $\frac{4}{9}, \frac{6}{7}, \frac{5}{3}, \frac{7}{4}$

61. $-\frac{7}{4}, -\frac{4}{9}, \frac{6}{7}, \frac{5}{3}$

3.7 Addition and Subtraction

1. $\frac{7}{5}$

3. 2

5. $-\frac{1}{3}$

7. $-\frac{5}{2}$

9. $-\frac{2}{3}$

11. $\frac{4}{3}$

13. $\frac{11}{8}$

15. $\frac{2}{4}$

17. $\frac{7}{4}$

19. $\frac{2}{9}$

21. $\frac{4}{35}$

23. $\frac{7}{8}$

25. $\frac{11}{12}$

27. $\frac{7}{36}$

29. $\frac{3}{2}$

31. $\frac{9}{40}$

33. $\frac{4}{5}$

35. $\frac{97}{180}$

37. $\frac{1}{3}$

39. $-\frac{1}{24}$

41. $\frac{7}{2}$

43. $\frac{13}{28}$

45. $-\frac{1}{2}$

47. $\frac{9}{8}$

49. $-\frac{7}{4}$

51. $\frac{17}{30}$

3.8 Solving Equations

1. 2

3. $-\frac{1}{2}$

5. 2

7. $\frac{1}{48}$

9. $-\frac{63}{5}$

11. $\frac{3}{2}$

13. $\frac{13}{2}$

15. $\frac{4}{15}$

17. $-\frac{35}{4}$

19. 6

21. 6

23. $-\frac{15}{4}$

25. $\frac{15}{2}$

27. $\frac{15}{64}$

29. $\frac{7}{8}$

31. $\frac{26}{5}$

33. $\frac{45}{8}$

35. $-\frac{11}{36}$

37. $\frac{1}{3}$

39. $\frac{14}{3}$

41. $-\frac{3}{7}$

43. $\frac{97}{90}$

45. $-\frac{11}{3}$

47. $\frac{121}{15}$

49. $\frac{163}{106}$

51. $\frac{2}{3}$

53. $\frac{3}{2}$

55. $-\frac{4}{5}$

Chapter 4: Expressions

4.1 Introduction to Expressions

1. Algebraic

3. Arithmetic

5. Algebraic

7. Algebraic

9. 343 m/s, three significant figures

11. 238,900 mi., four significant figures

13. 4,370,000 km, four significant figures

15. 300,000,000 m/s, three significant figures

17. 3.4×10^3 km

19. 1.989×10^{30} kg

21. 1.08321×10^{12} cu km

23. 8.6400×10^4 sec

25. 21 square feet

27. $\frac{16}{5}$ square inches

29. 21 square meters

31. 4 square miles

33. 98

35. $124,418.10

37. 140%

39. $899.33

41. 6.2 mi

43. 16 mm²

45. 16 km²

47. 41 ft²

49. 120 m²

51. 3 km²

53. 3 km²

55. 3 km²

57. 92 mi²

4.2 Combining Like Terms

1. $4x + 4$

3. $28x - 12$

5. $-2x - 4$

7. $t - 2$

9. $2g^2 + 5g$

11. $10x$

13. $10uv + 5u - 5v$

15. $2n^2 - 6n$

17. $7h^3 - 3h^2 - 2h$

19. $2u^3v^2 + 5u^2v^3 - 5u^2v^2 + 4u^2v$

21. $3x - 5$

23. $x - 38$

25. $-8n^2 + 18n + 24$

27. $2y^2 - 8y - 6$

29. $2(3x - 2)$

31. $4(w - 1)$

33. $x(4x - 5)$

35. $n^2(11n + 4)$

37. $4x(2x^2 - 3x + 5)$

39. $5dr(3d + 7r - 4)$

41. a) $a \cdot \frac{1}{c} + b \cdot \frac{1}{c}$

   b) $(a + b)\frac{1}{c}$

   c) $\frac{a + b}{c}$

43. $20m^2$

45. bh

47. $4k^2$

49. $17ab$

51. $26ab$

53. $\frac{3}{2}bh$

4.3 Simplifying Expressions with Algebraic Fractions

1. $\frac{2m}{3}$

3. $-\frac{4}{5}$

5. $\frac{2}{3}$

7. $-\frac{7t}{6}$

9. $\frac{8}{t}$

11. $\frac{3}{2k}$

13. $\frac{8y}{7}$

15. $\frac{2p}{3}$

17. $-\frac{2}{3s}$

19. $-\frac{m}{14}$

21. $\frac{1}{3y}$

23. $-\frac{v}{21}$

25. $\frac{9 - 2b^2}{15b}$

27. $\frac{14w^2 - 15}{18w}$

29. $\frac{21x + 55}{35x}$

31. $\frac{9 - 16a}{21a^2}$

33. $\frac{15p + 8}{36p^2}$

35. $\frac{4t^2 + 13}{2t(1 - 2t)}$

37. $\frac{5k^2 + 36}{15k(3k - 2)}$

39. $\frac{8w - 31}{9w(2w - 1)}$

41. $\frac{6m^2 + 7}{21m(m - 5)}$

43. $\frac{3}{4 - 12q}$

45. $\frac{3}{10x}$

4.4 Modeling with Expressions and Equations

1. Answers may vary. One possible answer is "Let g be the number of kilograms the giraffe weighs."

3. Answers may vary. One possible answer is "Let e be the number of graded exams."

5. Answers may vary. One possible answer is "Let l be the length of the rectange in mi."

7. Answers may vary. One possible answer is "Let i be the income of the California accountant in dollars per year."

9. Answers may vary. One possible answer is "Let E represent the number of extra dollars Larry earns each year."

11. $3 + a$

13. $d + 5$

15. $b + 4$

17. $m + 3$

19. $p - 6$

21. $5 - r$

23. $9 - t$

25. $5w$

27. $4y$

29. $\frac{a}{8}$

31. $2(3 + d)$

33. $1573 - Y$

35. $P + 3$

37. 81

39. 8

41. 7

43. 3

45. -56

47. -11

49. 2

51. $\frac{3}{2}$

53. 13 students

55. $5

57. 9 bricks

59. $770

61. 72 ft

63. 24.75 months

65. 12 years

4.5 Solving Equations

1. 5

3. $\frac{3}{4}$

5. $-\frac{17}{2}$

7. 4

9. $-\frac{3}{5}$

11. 3

13. -14

15. $-\frac{1}{3}$

17. $\frac{2}{3}$

19. $\frac{5}{6}$

21. 6

23. -6

25. no solution

27. no solution

29. -5

31. -1

33. 0

35. 2

37. $\frac{1}{2}$

39. $-\frac{1}{2}$

41. 2

43. -10

45. $\frac{11}{5}$

47. $-\frac{11}{9}$

49. no solution

Chapter 5: Graphs and Tables

5.1 Interpreting Graphical Representations of Data

1. 3.9 eggs per nest

3. 2003, 4.5 eggs per nest

5. 0.7 eggs per nest

7. Pizza, 20 students

9. 14 students

11. B

13. 80%

15. 70%

17. 10%

19. Carbon Dioxide, 21%

21. 100% All of something is 100% of it. If the percentages hadn't added to 100%, the difference would likely have been due to rounding the errors (rounding individual percentages up or down).

5.2 Mean, Median, and Mode

1. a) 5.6
   b) 6
   c) 8

3. a) 6.6
   b) 7
   c) no mode

5. a) 4.6
   b) 6
   c) 2 and 6

7. a) 5
   b) 4.5
   c) no mode

9. a) 4.8
   b) 4.5
   c) no mode

11. a) 6.3
    b) 7
    c) 8

13. a) 4.7
    b) 5
    c) 5

15. a) 6.3
    b) 7
    c) 5, 7 and 9

17. a) 5.5
    b) 6.5
    c) 8

19. a) 26.9
    b) 22
    c) 21 and 23

21. a) 745.6
    b) 837
    c) no mode

23. a) 72.4 in
    b) 71.5 in
    c) 70 in

25. a) 79 in
    b) 78 in
    c) 76 in

27. a) $4.2 \frac{g}{cm^3}$
    b) $2.2 \frac{g}{cm^3}$
    c) no mode

29. a) 210 frogs
    b) 210 frogs
    c) no mode

31. a) 71.6
    b) 70-79

33. a) 78.9
    b) 80-89

35. a) 36.1 yr
    b) 35-44 yr

37. a) 34.8 yr
    b) 25-34 yr

39. a) 73.6 sec
    b) 60-89 sec

41. a) mean 3, median 3, mode 3
    b) mean 19.2, median 3, mode 3
    Only the mean changed when an outlier was added, not the median or mode.

43. The agent would present the mean, because the outlier would bring the measurement up.

5.3 Equations with Two Variables

1. is a solution
3. is not a solution
5. is not a solution
7. is a solution
9. is a solution
11. is a solution
13. is not a solution
15. is a solution

17.

x	y
0	2
1	5
4	14
6	20

19.

x	y
−3	−2
−1	−4
0	−5
2	−7

21.

x	y
0	0
$\frac{1}{3}$	1
$\frac{2}{3}$	2
$\frac{3}{4}$	$\frac{9}{4}$

23.

x	Y
−2	5
0	1
$\frac{1}{2}$	0
2	−3

25. Answers may vary.
27. Answers may vary.
29. Answers may vary.
31. Answers may vary.
33. $x + y = 12$
35. $5x + 3y = 47$

5.4 The Cartesian Plane

1. Point A Q1
   Point B vertical axis
   Point C QIV
   Point D horizontal axis
   Point E QIII
   Point F origin
   Point G QII

3. Point A (3,1)
   Point B (0,3)
   Point C (−2, −3)
   Point D (0, −4)
   Point E (0,0)
   Point F (−4,1)
   Point G (4,0)

5.
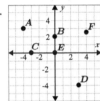

7. $-\frac{3}{2}$

9. 7

11. 0

13. 0

15. Point G

17. Point C

19. Point B (−2.0, −2.7)
    Point C (−1.2,0.8)
    Point D (0.4, −1.9)
    Point E (1.2,1.5)
    Point F (3.0, −1.1)
    Point G (3.3,1.0)

21.
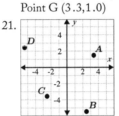

23. A. This scale is reasonable. It is okay to have different scales on different axes.
    B. This scale is not reasonable, because the length of a segment next to the origin is six units long, whereas the other lengths are one unit long. The segments must have uniform length horizontally and vertically.
    C. This scale is not reasonable, because the segments must have uniform length horizontally and vertically.
    D. This scale is reasonable. It is not necessary to label every mark, as long as the scale can be determined by the measurements given.
    E. This scale is not reasonable. The origin must be (0,0).

5.5 Graphs of Linear Equations

1. not linear

3. linear

5. not linear

7. not linear

9. $x - 2y = 8$

11. $x + 2y + 3$

13. $8x + 9y = 42$

15. $3x = 5$

17.

x	y
−2	−5
0	−1
2	−3

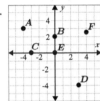

19.

x	y
−3	−1
0	1
3	3

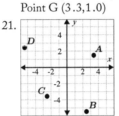

21.

x	y
−4	$-\frac{5}{3}$
−2	$-\frac{1}{3}$
2	$\frac{7}{3}$

23.

x	y
−3	3
0	4
3	5

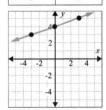

25.

x	y
0	5
2	1
4	−3

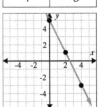

27.

x	y
1	3
3	-1
5	-5

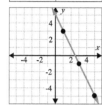

29.

x	y
-2	-3
0	-3
2	-3

31.

x	y
5	-2
5	0
5	2

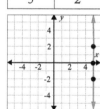

33. Problem 19 is easier to work, because the x-coordinates in the table make evaluating the y-coordinates easy. Choose your x-coordinates wisely, when possible.

35. Problem 27 is easier to work, because simplifying the equation before completing the table makes it easier.

37.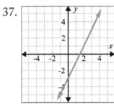

39.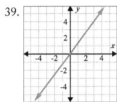

41.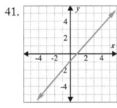

43.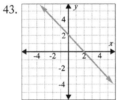

45.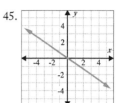

47.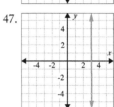

49.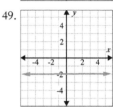

51.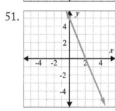

53.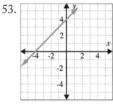

55.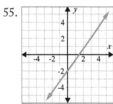

Chapter 6: Proportional Reasoning

6.1 Ratios and Rates

1. $\frac{29}{400}$

3. $\frac{3}{13}$

5. $\frac{14}{125}$

7. $\frac{9}{2}$

9. The ratio of Crater Lake National Park visitors to Oregon Caves National Monument visitors was 9 to 1.

11. The ratio of the John Day Fossil Beds National Monument gross land area to that of the Lewis and Clark National Historical Park is 41 to 10.

13. The ratio of Lewis and Clark National Historical Park December 2016 visitors to total 2016 visitors was 3 to 100.

15. The ratio of Oregon Caves National Monument July 2016 traffic count to total 2016 traffic count was 18 to 100.

17. 6 miles

19. 60 millimeters

21. 165,000 miles

23. 31 centimeters

25. 5.2 centimeters

27. 70 square kilometers

29. 25 feet per second

31. 29 square millimeters

6.2 Solving Proportions

1. 6

3. $\frac{8}{3}$

5. $\frac{7}{10}$

7. 33

9. −1

11. 1.375

13. 22.1

15. −4

17. $-\frac{20}{19}$

19. 156391 adult Chinook salmon

21. 122,500 pounds per year

23. 3750 defective widgets

25. 28 test subjects

27. 20,000 vampires

29. 28 trout

31. $\frac{10}{9}$ cup of sugar

33. He has played Brand 92 times this year.

35. 17

37. 12

39. 10

41. 7.5

43. 12

45. $\frac{9}{2}$

47. $b = \frac{148}{9}, c = 20$

49. $b = \frac{380}{7}, k = \frac{76}{7}$

6.3 Dimensional Analysis

1. $\frac{5,890 \text{ m}}{40 \text{ ft}}$

3. $\frac{9.296 \times 10^7 \text{ mi}}{40 \text{ ft}}$

5. $\frac{1 \text{ lb}}{0.4 \text{ oz}}$

7. $\frac{80 \text{ lb}}{3 \text{ oz}}$

9. $\frac{1 \text{ lb}}{16 \text{ oz}}$ and $\frac{16 \text{ oz}}{1 \text{ lb}}$

11. $\frac{1 \text{ cm}^3}{1 \text{ mL}}$ and $\frac{1 \text{ mL}}{1 \text{ cm}^3}$

13. $\frac{5,280 \text{ ft}}{1 \text{ mi}}$ and $\frac{1 \text{ mi}}{5,280 \text{ ft}}$

15. $\frac{1 \text{ m}}{100 \text{ cm}}$ and $\frac{100 \text{ cm}}{1 \text{ m}}$

17. 5

19. 40

21. 51

23. 4

25. 23.6 cm

27. 19.6 in

29. 51.2 oz

31. 34.8 oz

33. 7.5 cm

35. 2.0 ft³

37. 5.9 ft

39. 11.8 L

41. 8.7 L

43. 3598.9 oz

45. 6.2 slugs

47. 1.227×10^{10}

49. 3.845×10^8 m

Glossary/Index

A

Addition: The process of bringing two numbers together to make a new total. The math symbol for addition is the plus sign (+).
2

Addition principle of equations: If a and b are any mathematical expressions, either algebraic or arithmetic, and c is any rational number, then the equations $a = b$ and $a + c = b + c$ are equivalent.
186

Addition property for equations: When you add the same number to both sides of an equation, the resulting equation has exactly the same solution(s) as the original equation.
71

Addition/subtraction fact family: The four equivalent equations that correspond to any addition or subtraction fact.
14

Additive identity property: For any number a, $a + 0 = 0 + a = a$.
60

Additive inverse property: For any number a, $a + (-a) = -a + a = 0$.
60

Algebraic expression: A combination of numbers and operation symbols that include at least one variable.
152

Algebraic expressions are equivalent: Two equations are equivalent if every solution of the first equation is also a solution of the second equation and if every solution of the second equation is also a solution of the first equation.
186

Algebraic fraction: A fraction in which either the numerator or denominator is an algebraic expression.
171

Anti-identity property: For any number a, $a \cdot (-1) = -1 \cdot a = -a$.
66

Approximate unit fraction: An approximate unit fraction is a quotient of approximately equal measurements. The value of an approximate unit fraction is approximately equal to one.
273, 274

Area: The number of units that fit onto a surface.
24

Arithmetic expressions: A combination of numbers and operation symbols.
148

Assignment statements: Equations of the form: Variable = number or number = variable.
12

Associative property: For any numbers a, b, and c, $(a + b) + c = a + (b + c)$ and $(a \cdot b) \cdot c = a \cdot (b \cdot c)$.
61, 62, 64, 66, 67

Associative property of addition: When a, b, and c represent any three numbers, then $(a + b) + c = a + (b + c)$.
3

Associative property of multiplication: When a, b and c are any three numbers, then $a \cdot (b \cdot c) = (a \cdot b) \cdot c$.
   24

Average: See mean.
   202

Axes: Number lines that are perpendicular to each other and cross at their zero points.
   225

B

Base: The lower number in the expression xn; indicates the identical factors being multiplied in a power.
   25

Bases: The horizontal measurement of the bottom of a geometric object; can also be the top if the bottom and top are parallel to each other.
   166

C

Canceling a common factor: The process of canceling common factors if they appear both above and below the fraction bar.
   171

Cartesian plane: A geometric model similar to the number line for ordered pairs.
   225, 226

Center: The fixed point in a circle.
   253

Circle: The set of points on a plane that are all the same distance from a fixed point.
   253

Circle ratio: The ratio of circumference to diameter; a constant represented by the Greek letter π ("pi").
   254

Circumference: The distance that one would have to travel along a circle to return to one's starting point without ever reversing direction.
   253, 254

Clearing the denominators: The process of multiplying both sides of the equation by the least common multiple of the denominators of all of the fractions in the equation which produces an equivalent equation that has no fractions.
   190

Collinearity: The term for when points on a graph could fit on a single line.
   233

Common multiple: A shared multiple between two numbers.
   97

Commutative property: For any numbers a and b, $a + b = b + a$ and $a \cdot b = b \cdot$ a.
   61, 62, 64, 66

Commutative property of addition: When a and b represent any two numbers, then $a + b = b + a$.
   3

Commutative property of multiplication: When a and b are any two numbers, then $a \cdot b = b \cdot$ a.
   24

Composite figure: The result of combining two or more simple figures.
   154

Constant factor: A number that comes before the multiplication sign in a term.
163

Coordinates: The numbers in an ordered pair; coordinates determine where a point is on the Cartesian plane.
227, 228, 230, 231

Cross-canceling: The process of reducing a product of fractions before multiplication by canceling common factors of the numerator and denominator across the multiplication symbol.
171

Cross-multiplication: The act of setting cross-products equal to each other.
260

Cross products: For any two fractions, a/b and c/d, the product created by multiplying the numerator of one by the denominator of the other, ad and cb. If $ad = cb$, then the fractions are equivalent.
110

D

Decimal fractions: Let a be an integer and n a whole number. Numbers in the form $a/(10^\wedge n)$ are decimal fractions and can be written using decimal notation.
88

Decimal point: A period placed between digits in decimal notation.
88

Defining the variable(s): The practice of identifying the unknown quantity or quantities involved in a given situation and assigning variables to represent them.
177

Diameter: The greatest possible distance between two points on a circle; twice the distance of the radius.
253, 254

Difference: When performing subtraction, the number of objects that remain in the collection.
5

Dimensional analysis: A method for converting units.
270

Distributive property of multiplication over addition: Given any three numbers a, b, and c, the expressions $a(b + c)$ and $ab + ac$ are equivalent.
161

Dividend: When performing division, the quantity being divided.
27

Divisibility: The quality of one number being divisible by another number.
91

Divisible: For any two numbers a and b, if the quotient of a/b is an integer, then a is divisible by b.
91

Divisor: When performing division, the quantity being used to divide the dividend.
27

E

Equation: A mathematical sentence stating that two expressions are equivalent, that they have the same value.
11

Equivalent addition: Let a and b represent any two integers: $a + (-b) = a - b$ and $a + b = a - (-b)$.
57

Equivalent equations: Equations that share the exact same solution(s).
12, 13

Equivalent expressions principle: States that if a and b are equivalent expressions, and c is any expression, then the equations $a = c$ and $b = c$ are equivalent.
186

Equivalent fractions: If a/b is fraction notation for a rational number and c is any non-zero integer, then a/b and ac/bc are equivalent fractions. If two fractions are both equivalent to a/b, then they are equivalent to each other.
107

Equivalent multiplication: In the set of rational numbers, division can be rewritten as multiplication by using reciprocals.
119, 120

Evaluate: Find the output value of an expression or equation given a specific input value.
2, 3

Evaluating an algebraic expression: Two algebraic expressions are equivalent if each expression evaluates to the same number as the other under any substitution of a number for the variable.
152

Even number: If a number n is divisible by 2, then n is an even number.
92

Exponent: The upper number in the expression x^n; records the number of identical factors being multiplied in a power.
22, 25, 26

F

Factorization: The process of finding divisors of an integer, n, to create an integer product that evaluates to n.
102

Factors: Quantities being multiplied.
23

First quadrant: The region above the horizontal axis and to the right of the vertical axis.
225

Formulas: A literal equation that relates two or more real-life quantities.
152

Fourth quadrant: The region below the horizontal axis and to the right of the vertical axis.
225

Fraction notation: Notation that uses fractions to represent a number.
86

Fundamental theorem of arithmetic: The theorem that states if n is any integer greater than 1, then there is one and only one factorization of n in which every factor is prime.
102

G

Graph: A visual representation of values.
195

Graph of the equation: The result of plotting every solution of an equation on the Cartesian plane.
235

Greater than: For two numbers a and b, if a is to the right of b on the number line, then a is greater than b, written $a > b$.
48

Greatest common factor: For two numbers a and b, the greatest common factor (GCF) is the largest number than can be factored out of both a and b. Can also be applied to a collection of numbers.
95

H

Height: The vertical distance from the bottom to the top of a geometric object.
152, 166

Histogram: A bar graph that displays a data set sorted into bins. Each bin is a range of values, and every number in the data set belongs in exactly one of the bins. The height of each bar is the number of data values that is sorted into the corresponding bin.
202, 203,

Horizontal line: A line parallel to the x-axis, written $y = k$ where k is a constant.
241

I

Identity property of addition: When a represents any number, then: $0 + a = a$ and $a + 0 = a$.
4

Identity property of multiplication: When a is any number, then $1 \cdot a = a$ and $a \cdot 1 = a$.
25

Improper fraction: A fraction with a numerator greater than its denominator.
124

Integers: Positive or negative whole numbers.
47

Inverse properties of addition and multiplication: Definition goes here.
186

Inverse property of addition: If r is any rational number then there exists a unique rational number $-r$ such that $r + (-r) = -r + r = 0$.
131

Inverse property of multiplication: If r is any rational number, and if $r \neq 0$, then there exists a unique rational number $1/r$ such that $r \cdot 1/r = 1/r \cdot r = 1$
118

L

Least common multiple: If a and b are two non-zero integers, then the least common multiple of a and b is the least positive integer that is a multiple of a and a multiple of b. We denote this with LCM(a, b).
97,

Less than: For two numbers a and b, if a is to the left of b on the number line, then a is less than b, written $a < b$.
48

Like terms: Terms whose variable components, including the exponent(s) on the variable(s), are identical.
164

Linear equation: An equation meeting the following requirements: 1) there is at least one variable, 2) none of the terms contain more than one variable, and 3) each variable has an exponent of 1. The graphical representation is a straight line.
233, 234

Linear equations: Let a, b and c be any three constants such that a and b are not both equal to zero. Then the equation $ax + by = c$ is a linear equation with variables x and y. In addition, any equation that is equivalent to this equation is also a linear equation with variable x and y.
233, 234

Lowest terms: A fraction written such that the numerator and denominator are relatively prime and the denominator is positive.
108, 109

M

Mean: The quotient of the sum of the numbers and the number of numbers in a set. Also called the average of a set of numbers.
202

Measurement: A number followed by a unit name.
270

Median: The middle of a list of numbers sorted from least to greatest. If there are an even number of numbers, the median is the average of the two middlemost numbers.
204

Minuend: When performing subtraction, the number of objects that is originally in a collection.
5

Mode: The mode is the data value that occurs most frequently within a data set. If there is a tie for most frequent data value, then each value in the tie is a mode. If every data value occurs only once, then there is no mode.
206

Multiplication/division fact family: The four equivalent equations that correspond to any multiplication or division fact.
41

Multiplication principle of equations: That if A and B are any mathematical expressions, either algebraic or arithmetic, and c is any rational number, then $A = B$ and $Ac = Bc$ are equivalent as long as .
136, 186

Multiplicative identity property: For any number a, $a \cdot 1 = 1 \cdot a = a$
66

Multiplicative inverses: Another term for reciprocals.
117

N

Net worth: Positive net worth is when all of the money someone has is greater than all of the money they owe. Negative net worth is when all of the money someone has is less than all of the money they owe.
56

Number line: A straight line that extends infinitely far in both directions. Evenly spaced ticks are marked out all along the line. Each tick corresponds to an integer.
48

O

Odd number: If a number, *n*, is not divisible by 2, then *n* is an odd number.
92

Opposites: The corresponding positive and negative integers.
48

Ordered pair: An ordered pair is a set of two number values that are substituted for the variables in a specified order.
213

Origin: The point where the *x*-axis and *y*-axis cross, $(0, 0)$.
225

P

Perimeter: The total distance around the outside edge of a closed geometric figure.
161

Polygon: A closed geometric figure with straight sides.
161

Product: The result of multiplying.
23

Proper fraction: A fraction with a numerator less than its denominator.
124

Proportion: An equation of two ratios.
259

Q

Quadrants: The four regions on a Cartesian plane divided by the two axes.
225

Quotient: When performing division, the quantity that results from dividing the dividend by the divisor.
27

R

Radius: The distance from the center to any point on the circle.
253, 254

Rate: A ratio that includes units.
252

Ratio: A quotient that may be used to compare two numbers that represent measurements, amounts, or quantities.
250

Real numbers: A set of numbers that includes every rational number, but also many other numbers that are not rational.
225

Reciprocals: Two real numbers whose product is 1; often written in the form *a* and 1/*a*.
117

Reduce: The process of finding an equivalent fraction written in lowest terms.
109

Relatively prime: When two integers only have the number 1 as a common factor.
96

S

Second quadrant: The region above the horizontal axis and to the left of the vertical axis.
225

Similar units: Different units of measurement for the same characteristic.
272

Simple figures: A basic geometric figure such as a line, triangle, or rectangle.
154

Simplifying: The process of adding like terms and reducing the number of operations in an expression.
162

Skip counting: A process of counting using groups of objects. Functions as the basis for multiplication.
22

Solution: If an equation has a variable, then any value assigned to the variable that makes the equation a true statement is a solution.
11

Square foot: An area unit using feet as measurement.
24, 31

Subtraction: The process of taking away items from a collection to make a new total. The math symbol used for addition is the minus sign (-).
2, 5, 6, 7, 10

Subtrahend: When performing subtraction, the number of objects being removed from the collection.
5, 6

Sum: The result of addition.
2

Symmetric property of equality: If a and b represent expressions, then $a = b$ and $b = a$ must be equivalent equation.
14

T

Terms: A number, a variable, or combination that are separated by addition, subtraction, or division.
163

Third quadrant: The region below the horizontal axis and to the left of the vertical axis.
225

Trapezoid: a four-sided polygon in which two of the sides are parallel to one another.
166

U

Unit fraction: A unit fraction is a quotient of equal measurements. The value of a unit fraction is exactly equal to one.
273

Unit price: The ratio of price to the amount purchased expressed as a rate.
252

Unit rate: A rate reduced such that the second measurement is 1.
252

V

Variable: A letter or symbol that represents a number whose value is unknown or whose value can change.
11

Variable factor: A variable that comes after the multiplication sign in a term.
163

Vertical line: A line parallel to the y-axis, written $x = k$ where k is a constant.
240, 241

W

Whole numbers: Whole numbers are symbols that are used to count objects in a collection: $\{0, 1, 2, 3, 4, ...\}$.
1

Window: The part of the Cartesian plane that you can see in a plot.
227

X

X-axis: A number line that is typically the horizontal number line on a coordinate plane.
227

X-intercept: The point where a graph crosses the x-axis.
239

Y

Y-axis: A number line that is typically the vertical number line on a coordinate plane.
227

Y-intercept: The point where a graph crosses the y-axis.
239

Z

Zero dividend principle: When a is any number except zero, then $0/a = 0$.
29

Zero product property: When a is any number, then $a \cdot 0 = 0$ and $0 \cdot a = 0$.
29

CPSIA information can be obtained
at www.ICGtesting.com
Printed in the USA
FSHW021713100919

9 781943 536566